Robust Modelling and Simulation

Idalia Flores De La Mota
Antoni Guasch · Miguel Mujica Mota
Miquel Angel Piera

Robust Modelling and Simulation

Integration of SIMIO with Coloured Petri Nets

Springer

Idalia Flores De La Mota
Posgrado de Ingeniería
Ciudad Universitaria
Mexico City
Mexico

Antoni Guasch
Dept. Enginyeria de Sistemes
Automàtica i Inf. Ind. (ESAII)
Barcelona
Spain

Miguel Mujica Mota
Faculty of Technology
Amsterdam University of Applied Sciences
Amsterdam, Noord-Holland
The Netherlands

Miquel Angel Piera
Bellaterra (Cerdanyola del Vallès)
Spain

ISBN 978-3-319-53320-9 ISBN 978-3-319-53321-6 (eBook)
DOI 10.1007/978-3-319-53321-6

Library of Congress Control Number: 2017930651

Printed on acid-free paper

This Springer imprint is published by Springer Nature
The registered company is Springer International Publishing AG
The registered company address is: Gewerbestrasse 11, 6330 Cham, Switzerland

I would like to thank my parents for showing me the consequences of the effort we make and specially Christina for being supportive all these years.

Dr. Miguel Mujica Mota

To my beloved mother.

Dr. Idalia Flores De La Mota

Acknowledgements

The authors would like to thank the Aviation Academy of the Amsterdam University of Applied Sciences for supporting the publication of this book and the National Autonomous University of Mexico for supporting the Spanish version of this book. For the English version we want to thank Ms. Lesley Clarke for her review and translation.

Contents

Authors and Contributors

About the Authors

Idalia Flores De La Mota is a full-time professor in the Engineering Postgraduate Programme at UNAM. She received her Ph.D. in Operations Research at the Faculty of Engineering of the UNAM. She graduated in Master's with honours and obtained the Gabino Barreda Medal for the best average of her generation. She also received the medal for outstanding female professor of the engineering faculty. She has been a referee and a member of various Academic Committees at CONACYT. She has been referee for journals such as JART of the Center of Applied Sciences and Technological Development, UNAM, SIMULATION, EJOR and IJSPM. She has been invited as a speaker to international conferences and participates in international graduate programs in Italy and India. Her research interests are in simulation and optimization of production and service systems.

Antoni Guasch is a full-time professor at the Polytechnic University of Catalonia (UPC) and specialized in modelling, simulation and optimization of processes. He has led more than 40 projects for the aerospace, nuclear, textile, steel, automotive, transportation, mobility, pharmaceutical and water management sectors. And he was a coordinator of 8 Spanish research projects and partner leader in two European projects. He has also participated in the organization of several Spanish and international conferences, such as the European Simulation Multiconference (1994), where he was a coordinator. Dr. Guasch is co-author of a book on modelling and simulation that is used for teaching in several Spanish cities. He is currently responsible for the simulation and optimization of industrial processes of inLab FIB (http://inlab.fib.upc.edu/) and collaborates with Agbar (http://www.agbar.es/) in the development of optimization algorithms for the management of agricultural irrigation. He is also co-leading the development of tooPath (www.toopath.com) which allows real-time monitoring and monitoring of high-speed trains running in Spain.

Miguel Mujica Mota is an associate professor at the Aviation Academy of the Amsterdam University of Applied Sciences in the Netherlands. He was previously the subdirector of the aviation studies at the Autonomous University of Barcelona. He holds a Ph.D. and a M.Sc. in industrial informatics from the Autonomous University of Barcelona and a Ph.D. and M.Sc. in operations research from the National University of Mexico, all obtained with the highest honours. He is vice chair of the Dutch Benelux Simulation Society and secretary of the EUROSIM federation of Simulation Societies (2016–2019). Dr. Mujica Mota has given several courses in modelling, simulation methodologies and optimization in different countries for industrial and academic audiences. He has participated in several international projects in which simulation and optimization were the key factors for the success of them. He is also a level I researcher of the Mexican National Research system (SNI) where he also participates as a scientific evaluator for Latin America. He is the co-author of three books and numerous scientific papers on simulation, operations research, aviation, manufacturing and logistics. His research interests lie in the use of simulation, modelling formalisms and heuristics for the optimization and performance analysis of aeronautical operations, manufacture and logistics.

Miquel Angel Piera is the director of LogiSim, a recognized research group on Modeling and Simulation of Complex Systems, and the former director for Technical Innovation Cluster. He is a full-time associate professor in the Telecommunication and System Engineering Department at Universitat Autònoma de Barcelona. He graduated with excellence from UAB in Computer Engineering (1988), obtained M.Sc. from University of Manchester Institute of Science and Technology in Control Engineering (1991) and got his Ph.D. in 1993. He is a member of the Editorial Board of 3 international journals, has been the General Chair of 5 International conferences and is an invited speaker in 8 international conferences. He has published more than 30 journal and chapters and more than 100 conference papers. Dr. Piera has been nominated for the Outstanding Professional Contribution Award 2013 from the Society for Computer Simulation and the William Sweet Smith Prize 2015 by the Institution of Mechanical Engineers (UK). At present, Dr. Piera is a scientific advisor of Aslogic ([http://www.aslogic.es<http://www.aslogic.es/]www.aslogic.es<http://www.aslogic.es/>), a company specialized in the development of decision support tools relying on simulation techniques with application to manufacturing, logistics and transport.

Contributors

Idalia Flores De La Mota Operations Research Department, National University of Mexico, Mexico City, Mexico

Jaume Figueras Universitat Politecnica de Catalunya, Barcelona, Spain

Antoni Guasch Petit Universitat Politecnica de Catalunya, Barcelona, Spain

Miguel Mujica Mota Amsterdam University of Applied Sciences, Amsterdam, The Netherlands

Mercedes Narciso Farias Universitat Autonoma de Barcelona, Bellaterra, Spain

Miquel Angel Piera Universitat Autonoma de Barcelona, Bellaterra, Spain

Introduction

The objective of this book was to illustrate in a clear and concise way the basic knowledge of modelling and simulation using Petri nets as a central formalism and its use in SIMIO as a simulation program for solving logistic problems. We think that the combination of both tools is the main contribution to this book. Translating the developed models using Petri nets to SIMIO is made through an equivalence mapping in such a way that the resulting models are more robust, reliable and easy to verify, validate and maintain.

A system, in general, can have many conceptual models associated with it, and all of them are valid. In this book, the modelling approach is Petri nets. We illustrate the use of Petri nets through the implementation of diverse academic examples; once the reader is familiar with them and how the equivalence in SIMIO has been performed, he will be able to model more complex systems.

The content of this book is considered as a supporting material for M.Sc. courses in complex systems, manufacturing analysis and modelling and simulation courses in different areas that range from business and operations management to operations research. SIMIO is a program that can be found in www.simio.com

The book is structured in the following way:

Chapter 1 is an introduction to the modelling of systems in general, and in particular to simulation using discrete-event systems in a complete cycle, the reader with experience in simulation can omit reading this chapter. In Chap. 2, a review of the statistical concepts is presented. It is required for the modelling and simulation of discrete-event systems examples that are presented in Chap. 5. Chapter 3 is an introduction to ordinary and coloured Petri nets. The coloured ones are presented following a notation formalized by Dr. Kurt Jensen, and this representation allows to translate in a concise way the developed models to models in SIMIO. Chapter 4 introduces the elements used in SIMIO and coloured Petri nets for making an integration of both approaches. At the end of the chapter, two examples are presented and discussed. Chapter 5 presents some examples with the objective of illustrating another way of mapping the elements of Petri nets in SIMIO. Some of the examples are classic problems taken from the literature.

Chapter 1
Introduction to Digital Simulation

Idalia Flores De La Mota, Antoni Guasch and Miquel Angel Piera

1.1 Introduction

A computer simulation is an attempt to model a process from a real or hypothetical system by means of a computer program in order to observe, analyze and improve its behavior. In more practical terms, simulation can be used to forecast the future behavior of a system and determine what can be done to influence this behavior. If we want to analyze, study and improve any system by using digital simulation techniques, we must first develop a conceptual model that describes the dynamic in question and then encode it in a simulator to analyze the results.

On the one hand, simulation is in itself a very old technique that is inherent to the human learning process, as can be seen in children's games, which could be considered a simulation of the real world. On the other hand, digital simulation is a recent phenomenon as, in order to be able to understand reality and all the complexity that a system may imply, we have found it necessary to build artificial objects and dynamically experiment with them before interacting with the real system. Digital simulation can be viewed as the electronic equivalent of this type of experimentation.

Recently these techniques have become increasingly relevant in the solution of different types of practical problems; applications are now commonly found in fields such as engineering, economics, medicine, biology, as well as in ecology and social sciences. In fact, the development of mathematical models and how to run them in digital simulators is taught in a number of different university programs.

1.2 Definition of Simulation

What is simulation? Intuitively we can say that it is representing, pretending, acting. It is not much different from this in science, industry and education: simulation is a research or teaching technique that reproduces a similar or approximate form of real

I.F. De La Mota et al., *Robust Modelling and Simulation*,
DOI 10.1007/978-3-319-53321-6_1

events and processes them under certain predefined test conditions. Developing simulations of this type requires mathematical processes that are, in some cases, quite complex. Initially, a set of rules, relations and operating procedures must be specified. The interaction of these phenomena creates new situations or new rules that evolve as the simulation is developed.

Simulations can range from very simple ones that can be performed with a pencil and paper to sophisticated computer representations with interactive systems for almost real environments.

The origin of the modern use of the term *simulation* dates back to the paper of John Von Newman and Stanislaw Ulam at the end of 1940, when they constructed the term *Monte Carlo Method* for a mathematical technique used to solve problems in nuclear science that were either very expensive for an experimental solution or too complicated to be treated analytically. Historically, the term Monte Carlo was a code name that was used in the Second World War for secret calculations to predict the flow of neutrons in an atom bomb. The flow of millions of neutrons following random routes through a mass of uranium molecules can only be modeled on a computer, as it is impossible to forecast said flow theoretically. As the neutron routes vary at random and the building of the atom bomb was a huge gamble, the calculations were given the code name Monte Carlo, after the capital of the principality of Monaco, the most famous gambling center in the world.

Simulation took on a whole new meaning with the advent of computers in the nineteen fifties, as people could now experiment with mathematical models that represent a system. This made it possible to find a quick solution for problems that would have taken much too long to solve by hand. And, for the first time, social and administrative scientists found that they too, like the technicians, could perform controlled laboratory experiments. A series of new applications in every field quickly ensued. A suitable and up-to-date definition of simulation might therefore be:

Definition 1.1
Simulation is a numerical technique for performing experiments on a digital computer, making use of graphics, animation and other technological devices; that involves certain types of mathematical and logical models that describe the behavior of a system (or any of its components) during a particular time.

1.3 When to Use Simulation

Simulation is one of the most frequently used techniques and everything indicates that its popularity is on the rise. To analyze the reasons for its use, it is worth exploring the existing alternatives to simulation, in other words, the different methods that can be used to solve the same problem:

1. Direct experimentation with the real system or with a physical model of the real system.
2. Use of any other type of analytical mathematical model.
3. Use of experimentation by simulation.

Figure 1.1 shows a decision diagram for the use of models.

Simulation is used in most cases when the mathematical alternatives are poor, in other words, it is the "last resort", something like: "when all else fails, use simulation".

In reality, if the analytical solution is relatively simple, it will always be preferable to simulation, as it considers the general model. However, the problem is that there are a lot of systems that do not generate problems that can simply solved, so then we resort to simulation. For example, there are queuing problems that involve random processes distributed in a series of components of the system: inventory models, models for shared resources, time series forecasting, economic behavior, production approaches, vehicular movement, crossroads dynamics, etc.

Another advantage of simulation is that we can experiment without exposing the organization to the losses resulting from mistakes made in the real world. For example, some banks have studied changing their system from multiple queues to a single queue by employing simulation models and without needing to experiment on the customers. Direct experimentation with customers could have disagreeable consequences if it does not work as expected.

Furthermore, it is easier to control experimental conditions in a simulation model than in a real system. Consider the model of a traffic junction, where different synchronizations of traffic lights can be analyzed without affecting the real elements, which would be much too expensive in terms of money and human life.

A simulation model makes it possible to compress long periods of time and immediately analyze their behavior. We can, for example, visualize how the population will be in 30 years time and whether the transport services would be sufficient to meet its transport needs.

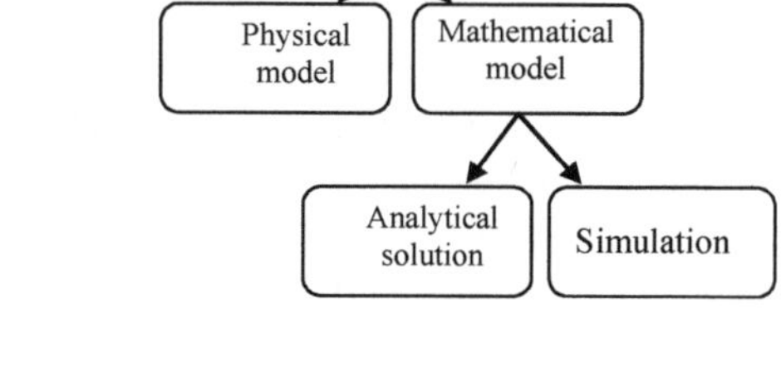

Fig. 1.1 Diagram for the use of models

Of course sometimes the system that we want to analyze does not even exist, in which case, simulation or a qualitative method would undoubtedly be ideal. Simulation does not, per se, replace other forms of experimentation or subjective judgment, but is a convenient alternative solution when the model is highly complex or the number of variables very large. In any case, experience and intuition, as well as a profound knowledge of the phenomena are essential ingredients for the success of our simulation models.

1.4 Concepts of Systems

In order to introduce the concept of modeling a system, as well as to present the different types of simulation models and their main characteristics, we must first define what we understand by the word "system".

Systemic analysis is a useful tool, given its tendency to study systems as entities rather than as a conglomerate of parts (Negroe 2005). This is in keeping with current scientific trends of not isolating phenomena into narrow contexts but rather examining the interaction between phenomena and thus studying increasingly larger parts of the world we live in. Moreover, systemic analysis, in the case of organizations, tries to get an overall view, in other words, to consider the system as a whole, i.e., it is not interested in solving a specific problem that someone thinks they might have, but rather it aims at looking at the entire organization in order that, after having conceived it as a system, it may proceed to study specific problems. According to Ackoff, the systemic approach observes the problems of systems as a whole; it is interested in the total performance of the system because, even when there are only changes in some of its parts, there are certain properties that can only be dealt with from a holistic perspective.

The relevant literature contains an abundance of definitions of system; one is to treat it as a defined aggregate of thoughts, concepts, judgments, mathematical relationships and logical connectives, whose unity and integrity are conditional on the interrelationships with the properties, bonds and links of the initial object. This means that the notion of a system, which is expressed with the help of signs, sentences in natural language, material means and technical constructions is nothing but a representation of the object under study (Negroe 2005). For our purposes, it is more important to have a definition that is an explicit process for the conceptualization of the system through its construction rather than the traditional descriptive definition.

Two basic types of systemic construction procedures have been defined (Negroe 2005): the composition procedure and the decomposition procedure; both are partial, complementary and produce two types of system representations: composite and integrated. The concept of *general system* is determined to be a constructor that is obtained with the composition of both representations.

1.4.1 Construction by Composition

This procedure commences with the initial attempts to define "system", which corresponds to the first stages of elaborating the concept, when one starts to understand that the set of elements selected is organized and interconnected in a certain whole and governed by common laws (Negroe 2005). In the next stage, the construction of the concept consists of attempting to deduce the properties of the system by studying its basic components, which are classified, and then finding the types of relationships that link them to each other. With this procedure, which takes the element as a starting point aiming to arrive at the system, we run the risk of not understanding its holistic nature; in other words, those aspects governed by the role it plays in a larger system known as a *suprasystem*. Thus, in these types of constructions, the set of elements, links and interrelations constitute one of the possible partial notions of the system, see Fig. 1.2.

1.4.2 Construction by Decomposition

This type of procedure is closer to the spirit of being systemic, corresponding, as it does, to a cognitive movement in the opposite direction to the above construction; in this case, we go from the system to its components, and this constitutes a typically holistic approach. The procedure is based on functional decomposition (widely used in I.T.); this consists of dismembering the system into subsystems, whose functions and properties, by proper organization, ensure those of the system as a whole. This approach is the one that will be used throughout this book, as the construction of simulation models requires us to be able to identify the parts of the system and their functions, see Fig. 1.3.

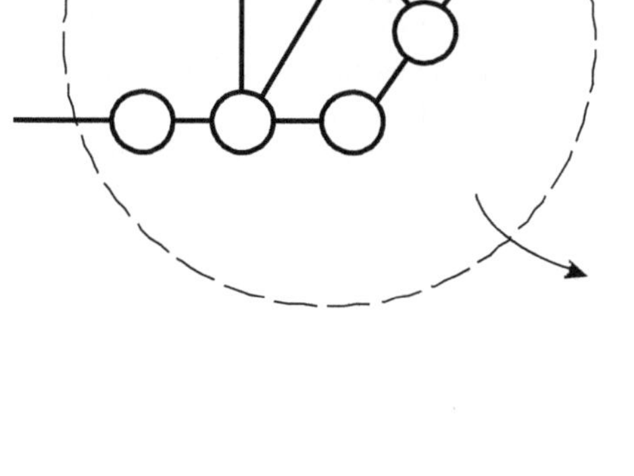

Fig. 1.2 Composed system representation through composition process

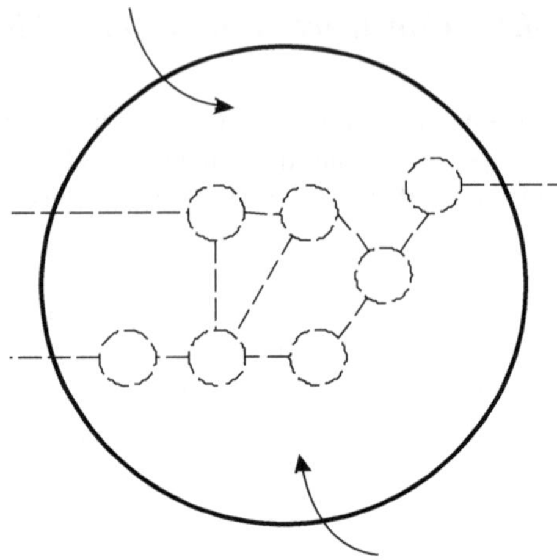

Fig. 1.3 Complete representation of the system through decomposition process

It must be pointed out that both the composition procedure and the decomposition procedure constitute partial and complementary notions that lead to the notion of the system as seen in Fig. 1.4.

1.4.3 Definition of System

A system can be defined as a set of objects or *entities* that interact to achieve a specific objective. If we plan to study the number of cashiers available in a bank to provide good customer service, the entities of the system will consist of the customers who are waiting to be processed and the cashiers responsible for providing the service.

1.4.4 State of a System

The *state of a system* can be defined as the minimum set of variables necessary to characterize or describe all those aspects of interest in the system in any specific instant of time. We call these variables *state variables*. It is useful to reiterate that the number and type of these variables depend on the objectives of our study. Then, in the example of a system described above, the state variables can be the state of each one of the cashiers (available or occupied in this case), and the total number of customers in the bank.

Considering the relationship between the evolution of the state variables and the independent variable *time*, the systems can be classified into continuous and discrete.

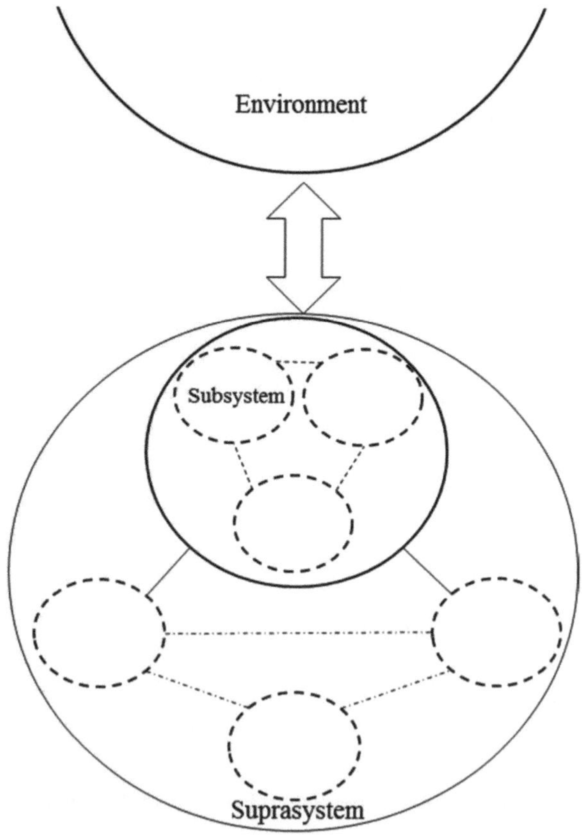

Fig. 1.4 Relationship between suprasystem, system, subsystem and the environment

- *Continuous systems*: the state variables continuously vary over time. One example of these types of systems is the evolution of the temperature in a room for any period of time, or the change in the level of a liquid in a tank, see Fig. 1.5.
- *Discrete systems*: so called because the system's state variables change in a particular instant or in a sequence of instants, and remain constant for the rest of the time. The sequence of instants in which the state of the system can present a change, normally follows a periodic or random pattern. If the sequence of instants follows a random pattern, the system can also be called a *Discrete event system*.

Example 1.1 Toll Road

One known process that illustrates discrete behavior is the toll road, where cars randomly arrive and where you want to foresee the waiting time of the cars in the lines that form at the tollbooths, so that said times may be considered as a function of the number of open tollbooths. Considering, for example, the changes in the state of a tollbooth (available, occupied), we can observe that the instants of time when

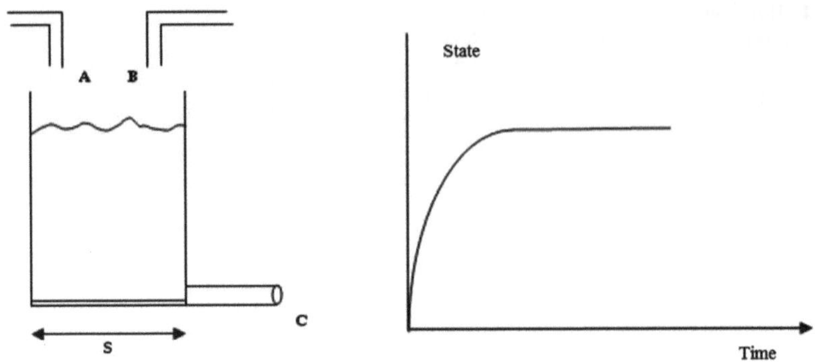

Fig. 1.5 Evolution of a continuous system

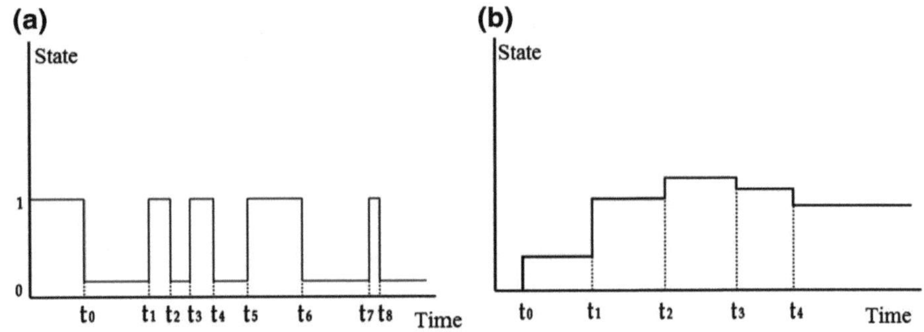

Fig. 1.6 **a** State of a tollbooth. **b** Evolution of the control signal in a constant sampling period

the change of state appear, follow a random pattern. Figure 1.6a represents the state of one of the tollbooths on a toll road, using value 1 to indicate that the tollbooth is available, and value 0 to show that the tollbooth is occupied. In contrast, the same Fig. 1.6b represents the evolution of the control signal logged by a computer with a constant sampling period.

1.5 Types of Models

There are other alternatives to digital simulation for imitating the behavior of a system. Some of the typical alternatives are the construction of a scaled prototype of the real system; the analogous representation of the system by electrical circuits; or the analogy with other biological or physical systems, such as experimenting with a new drug on animals to foresee its effects on human beings.

Many of these techniques have in common that, in order to imitate the behavior of a system, they need to have a description of its internal characteristics. The description of a system's characteristics of interest is known as a *system model*, and the process of abstraction to obtain this description is called modeling.

There are several types of models (physical, mental, symbolic) for representing the systems in question. However, as their use in digital simulation environments is one of the reasons for which the models are to be developed, the models need to formalize the knowledge about the system concisely and unambiguously - in other words, they must have a unique interpretation that can be processed by a computer. These characteristics determine the use of symbolic mathematical models as a tool for representing the dynamics of interest in any system in a digital simulation environment. Symbolic mathematical models map the relationships among the physical properties of the system being modeled in the corresponding mathematical structures. The type of mathematical formalization used is going to depend on the intrinsic characteristics of the particular dynamics that are to be represented.

Although there are a wide variety of methodologies for developing mathematical models of physical systems, the following considerations must be taken into account to guarantee an efficient representation of the real system:

- A model is always developed after a series of approximations and hypotheses and, consequently, only partially represents reality.
- A model is built for a specific purpose and must be formulated with the aim of it being useful for that purpose.
- A model must necessarily be a compromise between simplicity and the need to bring together all the essential aspects of the system under study.

Therefore, a good model must preserve two properties:

- Properly representing those characteristics of the system that we are interested in.
- Being an abstract representation of reality that is sufficiently simple to facilitate its maintenance, adaptation and reutilization.

In general, the complexity of modeling systems involves finding a balance between simplicity and precision in the model. *Ockham's razor* is a principle that is particularly relevant for modeling; its essential idea is that between models with approximately the same predictive power, the simplest is the most desirable. Although adding complexity generally improves the adjustment of a model, it can also make it harder to understand and work with, as well as imposing computational problems. Depending on the aims of the study and bearing in mind the characteristics that every good model must possess, simulation models can be classified and compared as follows:

1.5.1 Static Models Versus Dynamic Models

Static models generally represent the system in a particular instant and do not consider progress over time in their formulation. By way of illustration, a simple mathematical model, for example, considers the quantity of material in a warehouse of a particular factory:

Inventory = *Initial inventory* + *Incoming material* − *Material consumed by the factory*

These types of models are very useful when the system is in equilibrium (does not change over time). If the point of equilibrium is altered by changing one or more values of the system, the model lets us deduce the rest of the values, but does not show how they changed.

In contrast to static models, dynamic models allow us to calculate the variables of interest and observe how they evolve over time. One example of a dynamic model is the movement of material in a warehouse that depends on the entry and exit flow. The following equations mathematically describe the evolution of the inventory assuming that the variables of interest change continually or discretely, respectively, where F_i and F_0 represent the warehouse's entry and exit flows.

$$\frac{dS(t)}{dt} = F_i(t) - F_0(t)$$

$$S(k+1) = S(k) + F_i(k) - F_0(k)$$

1.5.2 Deterministic Models Versus Stochastic (Probabilistic) Models

A deterministic model is a model where a set of state variables is only determined by parameters in the model and by groups of previous states of these variables. Therefore, the deterministic models perform in the same way for a given set of initial conditions. Conversely, in a *stochastic model,* randomness is present and the state variables are not described by unique values, but rather by probability distributions.

1.5.3 Continuous Models Versus Discrete Models

Continuous models are characterized by presenting the ongoing evolution of the variables in question. In general, ordinary differential equations are used for modeling the evolution of a variable over time, or partial differential equations are also used for modeling the evolution of the variable, but in terms of space.

Similarly to the definition of continuous models, *discrete models* are characterized by discretely representing the evolution of the variables in question.

It is important to bear in mind that, based on the above classification of models, it is possible describe a continuous system by means of a discrete model and vice versa. The decision to use a continuous or discrete model depends on the particular objectives of each study and not so much on the characteristics of the model. Thus, for example, it is possible to find flow models for cars on an expressway, where a continuous formulation has been chosen because the study's objectives are focused, for example, on assessing the evolution of traffic in the presence of an accident, where the movement of one particular car is totally unimportant.

Discrete event models are dynamic, stochastic and discrete, where the state variables change their value in non-periodic instants of time, without being governed by a clock. These instants correspond to the occurrence of an event. So an event is defined as the instantaneous action that can change the state of a model.

1.6 Advantages, Disadvantages and Risks of Simulation

Simulation is widely used and increasingly popular for studying complex systems. Its advantages have already been presented by Law and Kelton (2000):

- Most real-world complex systems with stochastic elements cannot be described exactly by a mathematical model that can be analytically assessed. In consequence, simulation is the only type of research possible.
- Simulation allows us to estimate the performance of an existing system by using a series of projected operating conditions.
- Proposed alternative designs of systems or operating policies can be compared through simulation; to observe which best suits a specific demand.
- You can have better control of the experimental conditions in simulation than would generally be possible if the system itself were to be experimented on.

In brief, simulation is a candidate and a safe technique for:

- Answering questions of the type "what would happen if we make this change in…"
- Contributing to the reduction of the risk inherent in the decision-making process.

On the other hand, some disadvantages are:

- Every run of a stochastic simulation model only produces estimations of the true characteristics of a model for a particular set of input parameters. So several independent runs of the model will probably be required for each set of input parameters being studied.
- Simulation models are often quite costly and take a long time to develop.
- The huge quantity of numbers produced by a simulation study or the persuasive impact of realistic simulation generally creates a tendency to trust the results of the study more than can be justified. If the model is not a valid representation of the system under study, the simulation results, regardless of how impressive they may seem, will not provide any useful information about the system. A simulation model cannot be better than the data used. This is when one remembers the maxim: "Garbage in, garbage out" [http://en.wikipedia.org/wiki/Garbage_in,_garbage_out].

When organizing a simulation study through its different phases, we should pay particular attention to a series of potential risks (De Vin et al. 2004; Law 1986; Maria 1997):

- Using simulation when an analytical solution is more appropriate.
- Not having a series of well-defined objectives at the start of the simulation study. Sometimes the project is misdescribed or started out being very ambiguous and simplifications or assumptions are made on the hop. This can result in the simulation becoming the end rather than the means.
- Simulation alone cannot solve problems. It provides administrators with potential solutions for solving problems, as it depends on the individuals who are responsible to effectively apply the proposed changes. For this reason, regular communication is important. This is essential for ensuring that the right problem is solved, to promote the credibility of the model and making sure that the potential solutions are implemented.
- Insufficient knowledge about simulation, probability and statistical methodologies. A significant percentage of people involved in simulation models are only trained to use one specific software package, which is not enough. A simulation analyst must be an expert in simulation methodologies (validating a model, selecting initial probability distributions, design and analysis of simulation experiments, etc.) as well as in probability and statistics (probability distributions, confidence intervals, etc).
- Inappropriate level of detail in the model. A very common risk for inexpert simulation analysts is to have an excessive level of detail in the model. It will seldom be necessary to model every aspect of the system in order to make effective decisions while this would also be unviable for reasons of time, money or computational constraints.
- Not collecting good data about the system. Simulation cannot generate accurate results when the input data are inaccurate. If someone is modeling an existing system, it is important to collect data about key random variables of the system.

This is often overlooked because of project time constraints or because the analyst does not realize that this is an important consideration.
- The construction of the model or the collection of data, generally, takes more time that planned, which leaves little time for proper validation and verification; this can lead to wrong conclusions about the correction of the model.
- Believing that the so-called "user-friendly" simulation packages require a significantly lower level of technical skills. Most real-world problems require some knowledge of programming. Moreover, the modeler still has to deal with the formulation of the problem, data collection and analysis, the validation of the model, the modeling of a random part of the system, design and analysis of the simulation experiments and project management in general. These activities require a significant amount of technical skill and experience.
- Failure to use animation. Animation is useful for communicating the essence of the model to decision-makers, debugging simulation programs and suggesting improvements to the operational procedures of a system.
- Replacing a probability distribution by its mean value or using an unsuitable distribution. One common (but unfortunate) practice in simulation is to present a source of randomness for a system by using its perceived mean value and not its corresponding probability distribution.
- Not properly analyzing the output data. A stochastic simulation model does not produce true performance measurements for the model, it only produces statistical estimates of them. A simulation analyst must properly choose the length of a simulation run, the warm-up time (if necessary), and the number of independent replications of the model (each one using different random numbers).
- The use of the model is broadened to deal with questions for which it was never designed or to extrapolate results over and above the model's original field of application. The initial series of objectives and the model simplification hypotheses must always be borne in mind when carrying out experiments.

1.7 Lifecycle of a Simulation Project

There is a consensus of opinion among the people involved in the development and maintenance of simulation models about simple models being preferable to complex ones. Despite this, in many projects, the models are generally large and complex. We must emphasize that too much complexity in models not only has an impact on computational performance, but also affects other aspects, such as the time required for the development of the model, its maintenance, verification and validation.

Although on would think that this is a very intuitive concept, there is no definition or measure of *complexity* that is accepted as a standard by the experts in the field. Some authors relate the complexity of the model, for example, with the "level of detail" and others, with "the generalization of the system". Some of the advantages of working with simple models are:

- They are easier to implement, validate and analyze.
- It is simpler, somewhat less "painful", to discard a simple model that, for example, has a design mistake than a complex model in which a considerable number of expert staff hours have been invested.
- It is easier to adapt a simple model than a complex model if the operating conditions or hypotheses change in the real system.
- The project's total lifecycle is generally shorter.

A simulation project is dynamic by nature. The results obtained while it is being developed expose new problems, as well as any limitations that are inherent to the system under study. This can force us to reconsider the initial orientation of the project. Furthermore, the customer's motivation can also change throughout the project, as a consequence of the results obtained or because of external factors. To succeed in such a dynamic environment we need to use a correct methodology.

Table 1.1 shows the phases of a simulation project. Although it may seem that the development of a simulation process is sequential, in reality that is not how it is. For example, if the simulation model obtained does not pass the validation phase (phase 5), it is possible that the conceptual model will have to be altered as will the simulation model.

1. *Formulating the problem*
   Specifying the objectives is one of the most important tasks in a simulation project. All the modeling and analysis activities must be based on the objectives. If they are not clear or specific enough, there is a danger of the problem not

Table 1.1 Phases of a simulation project

Phase	Description
1. Formulation of the problem	Definition of the problem and adjustment of the objectives
2. Design of the conceptual model	Specification of the elements of the system and its interactions considering the objectives of the problem
3. Data collection	Identification, collection and analysis of the necessary data for the study
4. Construction of the model	Construction of the simulation model based on the conceptual model and the collected data
5. Verification and validation	Verification that the behavior of the model agrees with the conceptual model and the collected data, checking that the simulation model represents the real system
6. Experimentation and analysis	Analysis of the simulation results for the purpose of detecting problems in the real system and recommending improvements
7. Documentation	Providing documentation about the study that has been carried out
8. Implementation	Putting the decisions taken with the support of the simulation study into practice

being properly focused and their being incapable of responding to the created expectations. Therefore, in the initial phase of any simulation project, the objectives need to be identified and formalized for them to be precise, reasonable, understandable and measurable. These objectives will serve as a guide during the project.

2. *Design of the conceptual model*
Once the objectives of the simulation project have been formulated, the temptation to start building the model immediately must be avoided. This option generally leads to simulation models that have a lot of lacunas and are hard to maintain. For this reason, it is advisable to formulate or specify the simulation model used at a higher level of abstraction (conceptual model) than the level of the simulation code. The conceptual model specifies the most important structural relations of the system we are attempting to simulate and, consequently, constitutes an instrument for dialog and coordination between the various departments or groups involved.
It also corresponds to this phase to specify what results or statistics we expect to obtain from the simulation model, in order to answer the questions formulated in the definition of objectives.
3. *Data collection*
In general, we recommend questioning all the available information and data. What is the source?, when was it obtained?, how was it collected?, is it right?, is there enough or too much data? An indispensable condition for getting good results is to have good data. Unfortunately, this is not possible in many cases. Even so, the questions posed often need to be answered and reasonable hypotheses have to be implemented, in collaboration with the end user. If the data are limited or their quality doubtful, it is advisable to be prudent when reaching conclusions on the basis of the simulation results. Even in the cases where there are problems with the data, the knowledge acquired and the results obtained in the simulation study is valuable information for decision-making.
4. *Construction of the model*
People frequently put more effort into building the model than solving the problem. Preparing a functional model mistakenly becomes the objective that is given more priority. Our primary motivation should be to seek to understand the problem and to find solutions. In order to advance at a faster pace in the achievement of these objectives, it is recommendable to first construct one or more simplified models that characterize the most essential parts of the system.
5. *Verification and validation*
The presumption of innocence—to be innocent until proved guilty—is legal rights that, in many modern nations, the accused have in criminal trials. This asserts that nobody should be considered guilty until they are found guilty by a court of law. On the contrary, in the field of simulation, experience recommends assuming that all the models are wrong unless proven otherwise. One of the main dangers of simulation is "forgetting the real world and unquestioningly accepting the results of the model". In order to reasonable sure that the

simulation model represents reality and, as a consequence, takes strategic and operational decisions based on the results, it is absolutely necessary to verify and validate the simulation model.

Verification consists of checking that the model is run correctly and in accordance with the specifications (conceptual model), while validation lies in checking that the theories, hypotheses and the assumptions are correct. If the process does not yet exist, it is necessary to contrast the results with experts in the process in question to examine whether the model is behaving as expected. Validation is a difficult task, as there is no standard way of solving validation problems (Dijkum et al. 1999). Although this book does not talk much about validation, we should not ignore the importance of this phase.

6. *Experimentation and analysis*

   The experimentation resides in carrying out tests with the model in order to make inferences that enable us to have greater security in making decisions. In this stage, techniques like lowering the variance or the design of experiments are often used.

   In general, the end results obtained using the model are not the most important added value of a simulation study. The most valuable result is the profound knowledge acquired in the analysis process that gives qualitative and quantitative pros and cons for the different design options in question.

7. *Documentation*

   It is important to keep an up-to-date document that shows the status of the project. Therefore, the document will evolve and mature in parallel with the simulation project. The objectives sought with the documentation are:

   (a) To show the status of the project at any given time. Thus, all the technical or executive personnel that is connected with the project has up-to-date information about the progresses that have been made.
   (b) To report about the entire project (final document).
   (c) To facilitate the reuse of the model in cases where a possible interest in its future use is to be expected.

   We recommend the documentation contain the following information: introduction, objectives, hypotheses, physical description of the system, description of the conceptual and simulation models, verification and validation, analysis of the experiments carried out and conclusions.

8. *Implementation*

   Making decisions as a result of a simulation study is known as implementation. It is very important for a simulation analyst to regularly interact with the proper administrators. If the administrator or decision-makers understand and agree with the model's assumptions, they are more inclined to accept the model as valid and use the results in their decision-making (Law 1986).

References

De Vin, L. J, Ng, A., Jägstam, A. N., & Karlsson, T. (2004). Manufacturing simulation: Good practice, pitfalls and advanced applications. In *IMC21 Conference* (pp. 156–163). Limerick, Ireland, Septiembre 2004.

Dijkum, C. V., Tombe, D. D., & Kuijk, E. V. (1999). *Validation of simulation models*. Amsterdam: SISWO Publication 403.

Law, A. M. (1986). Pitfalls in the simulation of manufacturing systems. In *Proceedings of the 1986 Winter Simulation Conference*. New York.

Law, A. M., & Kelton, W. D. (2000). *Simulation modeling and analysis* (3a ed.). Tucson, Arizona: McGraw-Hill.

Maria, A. (1997). Introduction to modeling and simulation. In *Proceedings of the 1997 Winter Simulation Conference.* New York.

Negroe, P. G. (2005). Papel de la planeación en el proceso de conducción. *Cuadernillo de Divulgación 6*. México: Facultad de Ingeniería, UNAM.

Chapter 2
Elements of Statistics for Simulation

Idalia Flores De La Mota and Antoni Guasch

2.1 Introduction

This chapter presents statistical concepts and definitions that are used when designing a simulation model. We start, in the first instance, by considering a conceptual model, then the need to verify the initial data for the model, followed, if necessary, by the data that can be adjusted to some probability distribution, where validating this adjustment also involves statistical concepts. Later, once the results are in, we consider the experiments that need to be done, as well as the replications, ending up with the analysis of the data obtained.

As described in the previous chapter, the construction of a simulation model requires data collection. This fundamental phase of constructing simulation models can, according to Trybula (1994), take between 10 and 40% of the time required for the study. Fortunately there have been a lot of studies looking for ways to shorten this time, and Skoogh and Johansson (2008) have proposed a methodology that we will discuss further on. So it is necessary to have enough of the required data in the shortest time possible, for which we need to have an idea of what type of data about the system under study are required. According to Robinson and Bhatia (1995), the data can be classified as shown in the following Table 2.1.

Category A data are very convenient as the only work involved is their analysis and validation. Category B data require an additional effort as they have to be collected during the simulation study. Lastly the category C data, which corresponds to an estimation of the data, require strategy as well as careful and scrupulous design in order to maintain the quality of the model.

For good data collection, once we have identified into which of the three categories they fall, we need to answer some questions, such as:

What information do we have? A common fault in simulation studies that are not well delimited during the planning stage is because of more data than necessary or than can validated by the available data being extracted from the simulation.

I.F. De La Mota et al., *Robust Modelling and Simulation*,
DOI 10.1007/978-3-319-53321-6_2

Table 2.1 Data classification for a simulation. *Source* Robinson and Bhatia (1995)

Type of data		
Category	Availability	Cases
Category A	Available	Automated recording systems, previous measured data
Category B	Not available but collectable	The system has not been studied previously and there are no records
Category C	Not available or collectable	New processes or equipment

Another problem that has to be solved is the reliability of our data, therefore some questions that can support this process are:

- What data do we need?
- How will we get these data?
- How long does each stage of data collection take, approximately?
- With what information and how will the simulation results be validated?
- What configurations of the model should be run?
- How many runs should there be and how large should they be?

To collect information about the system's structure and operating procedures, the following considerations are required:

- A single document or an interview with one person is not enough. It is essential for the simulation analyst to talk to as many experts in the system as necessary to get a full understanding of the system to be modeled.
- Part of the information provided will invariably be wrong. If a certain part of the system is particularly important, then at least two experts in the system will be needed.
- The system's operating procedures may not be formalized.

Considering all of the above and in accordance with the methodology proposed by Skoogh and Johansson (2008) which is presented in the following figure, we propose the following steps (Fig. 2.1):

The steps listed in the above paragraphs are:

1. Identifying and defining relevant parameters.
2. Specifying the requirements for accuracy.
3. Identifying the availability of the data.
4. Choosing methods for collecting data that are not available.
5. Were all the specified data found?
6. Creating a datasheet
7. Compiling the available data
8. Collecting the data that are not available
9. Preparing a statistical or empirical representation, as the case may be.
10. Is the representation sufficient?
11. Validating the representations of the data.

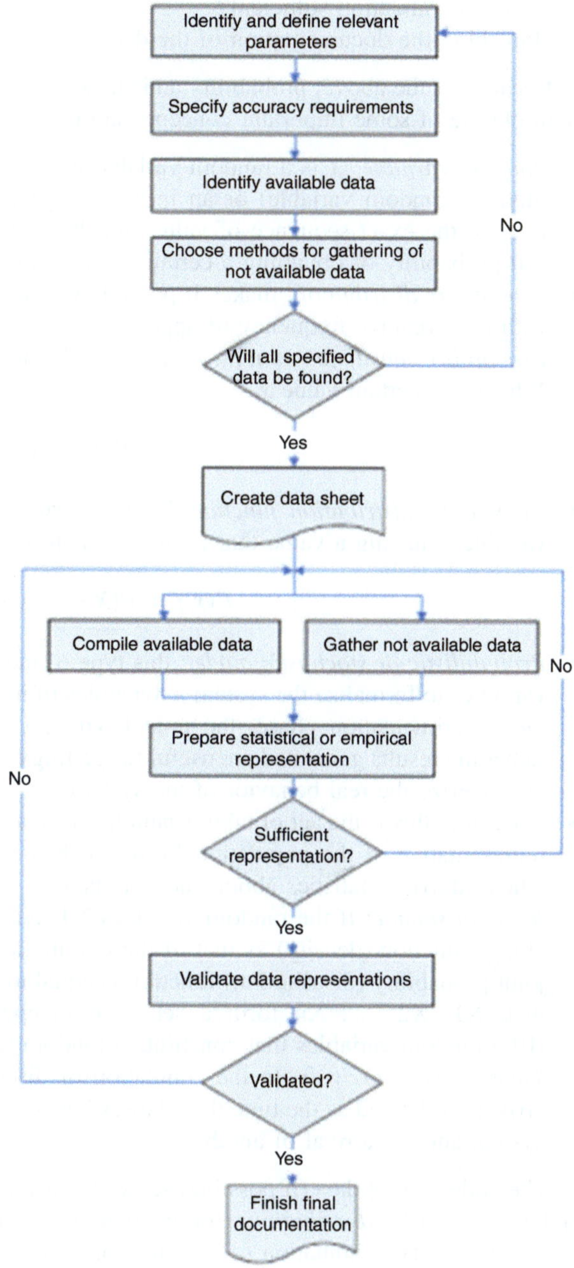

Fig. 2.1 Methodology proposed for increasing the accuracy and speed in the administration of the data for a simulation. *Source* Skoogh and Johansson (2008)

12. Was the validation sufficient?
13. Finishing the documentation of the data.

Because of the above, probability and statistical tools are indispensable, so here is an outline of some important concepts and definitions.

- *Random variable:* X is a random variable if it can take any value from a finite (discrete random variable) or an infinite range (continuous random variable). Despite the exact sequence of values not being known, the range of variation and probability of obtaining a certain value is known.
- *Probability distribution*: makes it possible to relate a set of values or measures with their relative frequency of appearance.
- *Probability density function* $f(x_i)$: describes the probability of a random variable X having a certain value x_i

$$f(x_i) = P(X = x_i)$$

- *Cumulative distribution function* $F(x_i)$: describes the probability of a random variable X having a value that is smaller than or equal to a certain value x_i

$$F(x_i) = P(X < = x_i)$$

- *Probabilistic or stochastic model*: this type of model uses one or more random variables to formalize the system's dynamics of interest. In consequence, during the experimentation phase, the model will not generate a single output, but rather the results generated are useful for getting estimations of the variables that characterize the real behavior of the system.
- *Sampling*: this is the act of taking samples from a population. A sample must be representative of the population from which it is obtained for it to be useful when inferring statistics about said population.
- *Random sample*: If the random variables X1, X2, …, Xn have the same probability function (density) as that of the distribution of the population and their joint probability (distribution) function is equal to the product of the marginals, then X1, X2, …, Xn form a set of n independent identically distributed (IID) random variables that constitute a random sample of the population.
- *Time between arrivals.* In the simulation of discrete events the time between arrivals is defined as the time that elapses between one arrival of an event to the system and the arrival of another event.

The collection of data (if possible) serves to specify the parameters of the model and the *probability distributions* (for example, for the failure time and repair time of the machine). The simulation of a system or process where there are components that are inherently random requires the generation of *random variables.* In the following sections we discuss how these values can be conveniently and efficiently generated from a desired probability distribution for their use in the simulation models. We also include the distribution functions that are more frequently employed in simulation models and the cases they are used in are specified. Care must be taken not to commit two common errors at this level: replacing a probability distribution by its mean value or using an unsuitable distribution.

All the discrete simulation packages that are available contain a random number generator however it is important to briefly discuss what this phase of the modeling consists of.

2.2 Generation of Random Numbers

Random simulation methods were initially applied by mathematicians and physicists to solve certain deterministic problems that could be expressed as mathematical equations whose solutions could not be easily obtained by the usual numerical or analytical methods. In many significant mathematical problems, we can find a stochastic process with a probability distribution or parameters that satisfy the requirements of the equations. The deterministic problems that stochastic simulation has been used for include the evaluation of multiple integrals, the solution of very high order differential equations, complex queuing problems and schedule programming. Although there are analytical methods for these cases, simulation methods have been found to be more effective.

Another type of problem that leads to the simulation of random variables arises in those situations where there is stochastic behavior and that require some type of sampling, which in practice is either impossible or inconvenient, as in the case of future data. Although we cannot get the data, we know something about the population from which it comes. For stochastic simulation, it will then be necessary to build a probabilistic model that is tailored to the study. This means that shall be indispensable identify one (or several) probability distribution(s) tailored to each case, which makes it possible to generate values that behave similarly to the phenomenon in question. Nowadays, most statistical and simulation packages include a random number generator, however we still cover the issue in this chapter as it a good idea to have a better understanding of what it means to generate these numbers.

The methodology for generating random numbers has a long and interesting history. The first methods were developed practically by hand, such as tossing coins, choosing cards, throwing dice or taking numbered balls out of an urn (Fig. 2.2). A lot of lotteries currently operate this way. At the start of the 20th Century, statisticians followed gamblers into their interest in random numbers and mechanical devices were built for a speedier generation of random numbers. In 1938, Kendall and Babington-Smith used a fast-spinning disk to prepare a table of 100,000 random digits. Sometime later they developed electrical circuits based on randomly pulsating vacuum tubes to throw out random numbers at a rate of 50 numbers a second. The Royal Mail used this type of machine: Electronic Random Number Indicator Equipment (ERNIE), to choose the winners of the Premium

Fig. 2.2 Selection of random numbers

Bonds lottery. The Rand Corporation used another similar device to generate a table of one million random digits.

Many other approaches have been used to randomly select numbers, such as the selection of numbers at random from the telephone directory or from census reports, or using digits taken from the decimal expansion of π.

As computers as well as simulation were being used more, there has been more interest in methods for generating random numbers that are compatible with the way computers work. Thus, research in the 1940s and 1950s focused on numerical or arithmetic ways of generating random numbers. Said methods are sequential, with every new number determined by one or more of its predecessors and, new ones are generated according to a mathematical formula, as we will see in some examples given in the following section.

2.3 Properties of a Good Random Number Generator

Because of the characteristics of simulation, it is necessary to generate random numbers that represent the behavior of the problem to be simulated. Although there are many ways of generating random numbers, for the majority of real applications the generator shall have a series of properties that make it truly useful and similar to the real processes:

- It must produce random numbers.
- It must be fast.
- It must have a long period before repeating its cycle.
- It must generate random numbers that can be reproduced.
- It must not require a lot of computer storage space.
- It must not degenerate.

Each of these properties are explained as follows:

In the production of random numbers, the fact that they are random means that they have to be independent of each other. They should initially come from a uniform distribution, which means that they are not strictly random, as their generation is based on a function but, for all practical purposes, this is what best fits the concept. This is the reason why they are called pseudo-random because, in reality, there are no cycles.

Large-scale simulation models generally require a lot of random numbers, so the generating method must be fast while the time and memory used in the computer must not be excessive.

In practical terms, all the generating methods produce numbers that sooner or later repeat their cycle at some point. This means that the sequence of numbers is repeated. Then, it is important for the generating method that is chosen to produce all the random necessary numbers before a cycle is completed. This suggests that the selection of the method will depend on the specific application. If 100 numbers are needed, a 200-long cycle will not cause any problems.

Also, it is important for the random number generating method not to degenerate, in other words, for the method not to repeat the same number indefinitely. For example, some methods degenerate at the zero value.

Consequently, we have to look for algorithmic procedures for the generation of number that are at least apparently random. Von Neumann's idea was to produce numbers that look random employing the computer's arithmetic operations. Starting from a seed or initial value (u_0, $u_{-1}, \ldots, u_{-p+1}$), a sequence is generated by means of $u_i = d\,(u_{i-1}, \ldots, u_{i-p})$ for a certain function d. With the seed having been chosen, the sequence is set.

The first question to be resolved is what do we mean by random numbers, which is why it is important have a good definition. Starting out from the modified version of Kolmogorov and Uspenskii's classic definition (1987) associated with the idea of algorithmic complexity, we have the following definition:

A sequence of numbers is random if it cannot be efficiently produced by a program that is shorter than the string itself.

For example, the sequence 0010010010… is interpreted as being non-random, given the fact that we can give a shorter algorithm than the string itself. The discussion of these ideas leads to interesting proposals. For example, a criterion for the definition of random numbers can be introduced that is similar to Turing's for recognizing an artificial intelligence, which brings us to the following definition:

A succession of numbers is random if nobody using reasonable computer resources in a reasonable time is able to distinguish between the series and a truly random sequence better than throwing a coin to decide which one it is.

The precise expression of this definition leads to the ones known as PT-perfect generators (Lécuyer 1990), of great interest in cryptography, but not in simulation, because of its slowness.

2.4 Generation of Random Numbers with a Uniform Distribution Between Zero and One

Discrete case

The importance of generating random numbers consists of them representing the value of a random variable; thus, if the variable is discrete and can only take n given values which are $x_{i,}$ $1 \leq i \leq n$, whose probability is $p_{i,}$ we know that:

$$\sum_{i=1}^{n} p_i = 1$$

These probabilities can be obtained in advance or else be determined through a series of observations, on which basis the different probabilities are established, as shown in the following example:

Example 2.1

At the junction of streets A and B, the following observation was made of the vehicles traveling along street A. Table 2.2 gives these observations:

The stochastic variable x_i can take one of the following three values: turning to the right, turning to the left or not turning. Observe that the possible values do not necessarily have to be numerical, they could, in this case, be actions. For it to be discrete it is necessary for the number of possible results to be finite, with the probabilities that are given in Table 2.3 (Fig. 2.3).

To generate random numbers with these probabilities, we consider the cumulative probability graph given in Fig. 2.4.

We generate a sequence r_i of random numbers with uniform distribution between zero and one and, depending on the range where the random number is found, this will be the value we associate with it, as shown in Table 2.4.

Congruential or residual methods

Congruential random number generating methods first occurred to Lehmer in 1951. These methods are based on what mathematicians call congruence relations. Although there are a lot of variants, the most popular ones are the multiplicative congruential generators or power residue methods.

This method, like the one above, requires a first number, after which a string of random numbers is generated by the recursive application of the following formula:

$$X_{i+1} \equiv \alpha X_i \quad (\text{module } m)$$

This relation is read as "X_{i+1} is congruent with αX_i in module m". By definition, two integers A and B are said to be congruent in module m (with integer m), if $(A-B)$ is divisible between m and if A and B produce identical residues when divided by m, this means that A is congruent with B in module m, if and only if there is a value k in the integers, so that

Table 2.2 Observations at the junction

x_i	Remarks	Probability (p_i)	Probability accumulated
To the right	28	0.28	0.28
To the left	23	0.23	0.51
Do not turn	49	0.49	1.00

Table 2.3 Probabilities associated with the junction

Value	Right	Left	Does not turn
Probability	0.28	0.23	0.49

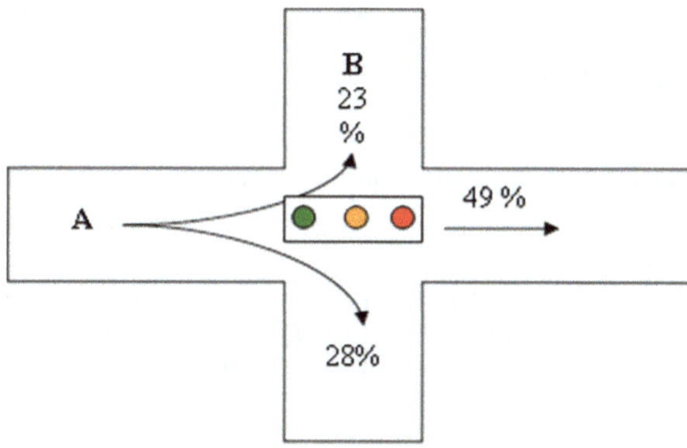

Fig. 2.3 Probabilities associated with the junction in the form of a graph

$(A-B) = km$.

Any sequence of numbers can be obtained by multiplying the preceding number by a constant and then reducing product by module m. The module m operation means dividing αX_i between m and keeping the residue as the value of X_{i+1}.

For example, let $\alpha = 5$, $X_0 = 3$ and $m = 32$. The value of X_1 will be 15 as:

$$X_1 \equiv (5)(3)(\text{mod } 32)$$

$$15/32 = 0 \text{ with } 15 \text{ as residue}$$

$X_2 = 11$ can be obtained in the same way. The distribution of the X_i is uniform and a source of random numbers.

Another way that is even simpler consists of using the "random" that different programming languages have. In the case of congruential methods, their length depends on the chosen module.

These are only some of the methods for generating of random numbers. There are many more and if the problem in question requires special treatment as regards randomness, it is important to consider the possibility of the random numbers being generated by the modeler or else through a programming language when the simulation is being executed and not using software that already includes it.

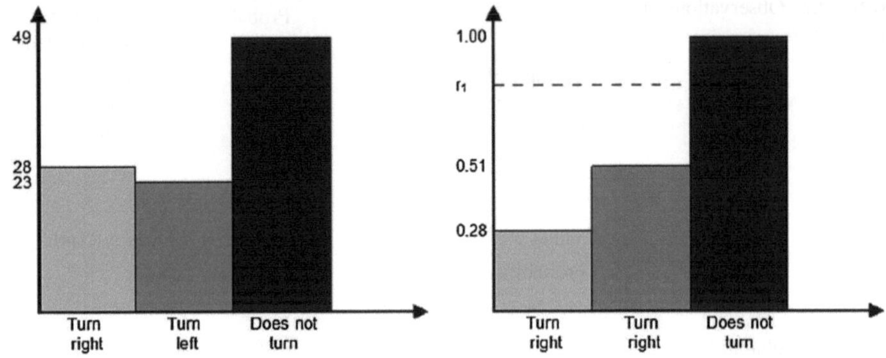

Fig. 2.4 Cumulative probability

Table 2.4 Ranges and cumulative probability

r_i range value of x_i	$i = 1$	$i = 2$	$i = 3$	$i = 4$	$i = 5$
	0.274 (0,0.28) right	0.911 (0.51,1) Does not turn	0.046 (0, 0.28) right	0.466 (0.28, 0.51) left	0.4976 (.0.28, 0.51) left

2.5 Selection of a Distribution Function

In some cases, the initial problem in the simulation of random variables is precisely the choice of a suitable distribution function. There are four considerations you may take into account for this selection:

1. *The special characteristics of each specific distribution.* This means the particular behavior that the phenomenon in question may have. For example, if the data has only two distinct values, the proper distribution will doubtless be a Bernoulli. If it is a question of discrete data, we can immediately leave to one side all the continuous distributions and vice versa. If it is a sampling, we should observe whether this is with or without replacement, in which case binomial or hypergeometric distribution, respectively, is used. Another important datum is the symmetry or asymmetry of the data of the phenomenon in question; for example, normal distribution is symmetric, whereas triangular distribution may or may not be. The times between events tend to be distributed as exponential, for continuous time.
2. *The accuracy with which a distribution can represent a set of experimental data.* This is only verified through graphs, such as the data histogram, which was obtained from the frequencies which were observed and goodness-of-fit tests.
3. *The facility with which the distribution fits the data, in other words, the estimation process for the corresponding parameters.* In some distributions, the process of obtaining estimators is extremely complicated and time-consuming,

particularly for models composed by equations with non-linear parameters. In these cases, we can resort to a simpler approximate distribution or make use of iterative algorithms, in order to get estimators that are sufficiently suitable for the particular needs of each problem.

4. *Computational efficiency in generating random variables.* As we have already mentioned, in some cases you have to do a lot of calculations to generate a set of variables. The simpler these calculations, the more efficient will be the calculation to obtain a large number of variables, which is important as large samples are desirable if we want to extract reliable conclusions.

The use of probability and statistics is an integral part of a simulation study and are used to understand how to model a probabilistic system that meets the following characteristics:

- To validate the simulation model.
- To choose the probability distributions to start with.
- To obtain random samples from the distributions.
- To make a statistical analysis of the simulation results.
- To design the simulation experiments.

There can be different probability distributions depending on the system and the problem to be solved, as shown in the following (Table 2.5):

2.6 Continuous Distribution Functions

A summary table is included for each one of the functions presented. This summary table includes the probability density function, the cumulative distribution function, the mean and the variance. The mean (or expected value) and the variance of a continuous random variable X that follows a density function, is calculated by means of:

$$\mu = \mathrm{E(X)} = \int_{-\infty}^{\infty} xf(x)dx$$

Table 2.5 Initial probability distributions

Type of system	Sources of chance
Manufacturing	Process times, time between breakdowns, arrivals of orders
Communications	Time between arrivals of messages, length, type, end destination
Transport	Size of the load, transport time, loading and unloading times
Hospital processes	Time between arrivals of patients, type of illness, length of consultation

$$\sigma^2 = V(X) = E(X - \mu)^2 = \int_{-\infty}^{\infty} (x - \mu)^2 f(x)dx$$

2.6.1 Exponential Distribution Function

This distribution is used to model the time between entities. It is also employed for modeling service times that are highly variable; for example, the length of time of a phone call. This distribution is related to Poisson distribution, given that if an arrivals rate (arrivals per unit of time = λ) follows a Poisson distribution, the time between arrivals follows an exponential distribution of parameter $\beta = 1/\lambda$.

Normal, log-normal and gamma distribution functions tend to be more frequently used for modeling those activities where, under normal operating conditions, the time consumed usually shows (physically justifiable) variations in respect of an average value (Fig. 2.5 and Table 2.6).

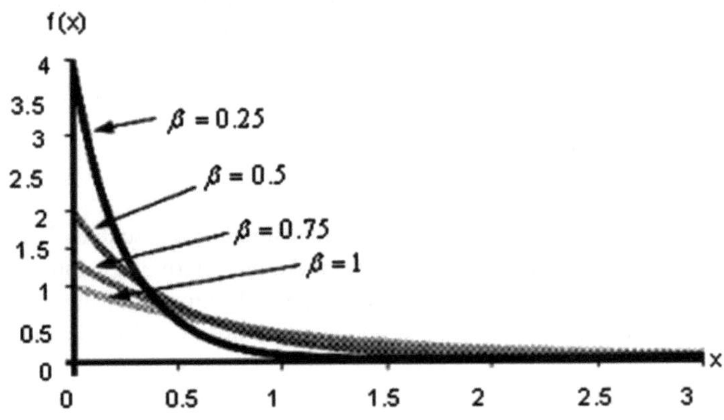

Fig. 2.5 Exponential distribution

Table 2.6 Exponential distribution function

Exponential	*Expo* (β)
Possible interest	Time between arrivals of customers when the mean frequency of arrivals is constant
Probability density	$f(x) = \begin{cases} \frac{1}{\beta} e^{\frac{-x}{\beta}} & x \geq 0 \\ 0 & x < 0 \end{cases}$
Cumulative distribution	$F(x) = \begin{cases} 1 - e^{\frac{-x}{\beta}} & x \geq 0 \\ 0 & x < 0 \end{cases}$
Mean	β
Variance	β^2

2.6.2 Gamma Distribution Function

In general, the time that a production unit requires to carry out a repetitive raw materials processing operation or the time consumed in a repetitive activity of transporting material between two work stations usually follows a constant value with small variations caused by certain physical aspects. These could be conclusively modeled but, in order to simplify the task, are usually described as the result of a random activity through statistical models.

In accordance with the parameters of the gamma probability distribution function (pdf), it shows a very similar graph to that of the normal pdf, but with a certain asymmetry that answers to the presence of data with values that are higher than the average value. This asymmetry makes it possible to model sequences of activities (for example, processing units or transport units) that are done in parallel, so that each one of them answers to a normal pdf, but the time consumed in the sequence of activities shows an asymmetry slanted towards the values that are higher than the average (Table 2.7).

Figure 2.6 shows different shapes of the gamma distribution in accordance with the variation of their parameters α and β, which are, respectively, shape and scale parameters.

The gamma distribution function represents a very good statistical modeling tool for modeling real systems submitted to the occurrence of certain events; for example, probability of machine failure, which increases the appearance of values higher than the average value.

Table 2.7 Gamma distribution function

Gamma	$Gamma(\alpha, \beta)$
Probability density	$f(x) = \begin{cases} \frac{\beta^{-\alpha} x^{\alpha-1} e^{\frac{-x}{\beta}}}{\Gamma(\alpha)} & x \geq 0 \\ 0 & x < 0 \end{cases}$ $\Gamma(\alpha)$ is the gamma function $\Gamma(\alpha) = \int t^{\alpha-1} e^{-t} dt$ If α is a positive integer $\Gamma(\alpha) = (\alpha - 1)!$
Cumulative distribution	$F(x) = \begin{cases} 1 - e^{-\frac{x}{\beta}} \sum_{j=0}^{\alpha-1} \frac{(x/\beta)^j}{j!} & x \geq 0 \\ 0 & x < 0 \end{cases}$ If α is a positive integer; otherwise there is no closed formula
Mean	$\alpha\beta$
Variance	$\alpha\beta^2$

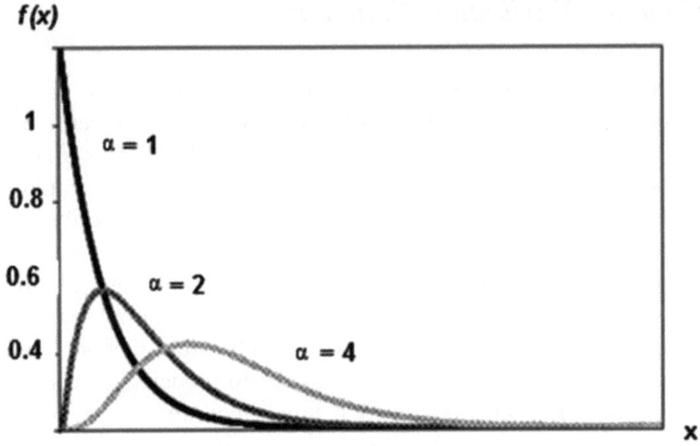

Fig. 2.6 Gamma distribution function ($\beta = 1$)

2.6.3 Log-Normal Distribution Function

In general, the log-normal distribution function is used for modeling a multiplicative sequence of operations; for example, the repercussion of the failure of one machine is therest of the machines being shutdown. The gamma distribution function is used for modeling an additive sequence of operations, while the log-normal distribution function can be used for modeling the time required to do a manual task (Fig. 2.7 and Table 2.8).

2.6.4 Normal Distribution Function

This is used for modeling systems where 70% of the sampled data is found at a distance of less than σ (standard deviation) from the average value μ, and the frequency of appearance of the data is found symmetrically distributed in respect of the average value.

One example for using a normal distribution function is the modeling of the production time of the machines, when the possibility of different types of faults or errors is not considered.

Figure 2.8 represents the histogram of a normal distribution function, in which the difference of the gamma and log-normal pdfs, the data practically does not present huge variations in respect of an average value (Table 2.9).

The cumulative distribution function cannot be accurately calculated. As a consequence, numerical methods were employed to obtain tables for the function. Given that it is not practical to obtain a table for all the possible values of μ and σ^2, a table is constructed for the standard normal distribution (of parameters $\mu = 0$,

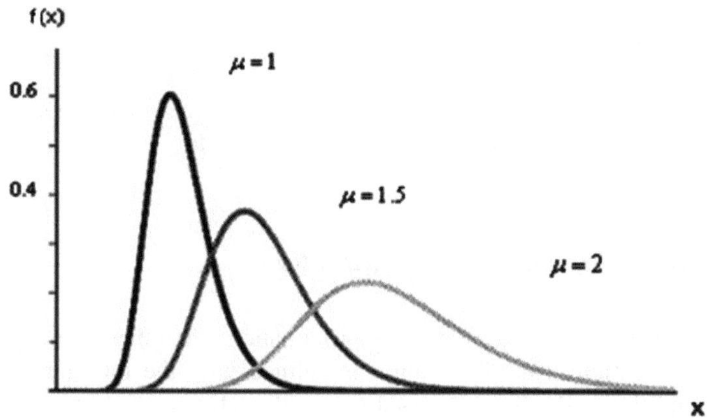

Fig. 2.7 Log-normal function σ = 1

Table 2.8 Log-normal distribution function

Log-normal	$LN(\mu, \sigma^2)$
Probability density	$f(x) = \begin{cases} \frac{1}{x\sqrt{2\pi\sigma^2}} \exp\left(\frac{-(\ln x-\mu)^2}{2\sigma^2}\right) & x \geq 0 \\ 0 & x < 0 \end{cases}$
Cumulative distribution	There is no closed formula
Mean	$e^{\mu+\sigma^2/2}$
Variance	$e^{2\mu+\mu^2}(e^{\sigma^2}-1)$

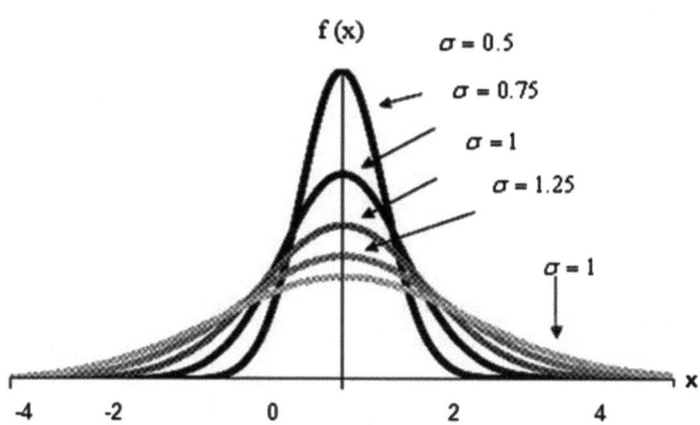

Fig. 2.8 Normal distribution function (μ = 0)

σ^2 = 1). If X is a random variable of normal distribution of values of, μ and σ^2 the random variable $Z = (X-\mu)/\mu$ follows a normal distribution of mean 0 and variance 1.

Table 2.9 Normal distribution function

Normal	$N(\mu, \sigma^2)$
Probability density	$f(x) = \frac{1}{\sqrt{2\pi\sigma^2}} e^{\frac{-(x-\mu)^2}{2\sigma^2}}$
Cumulative distribution	There is no closed formula
Mean	μ
Variance	σ^2

2.6.5 *Triangular Distribution Function*

Triangular distribution provides a first approximation when there is not very much available information. This distribution is defined with the minimum value, the maximum and the mode. It is also used to specify activities that have a minimum, maximum and more probable time (Fig. 2.9 and Table 2.10).

2.6.6 *Uniform Distribution Function*

The uniform distribution is a continuous distribution that is used to specify a random variable, which has the same probability of having its value at any point on a range of values. It is defined by specifying a lower bound and an upper bound b for the range. Uniform distribution is not, in general, a valid representation of a

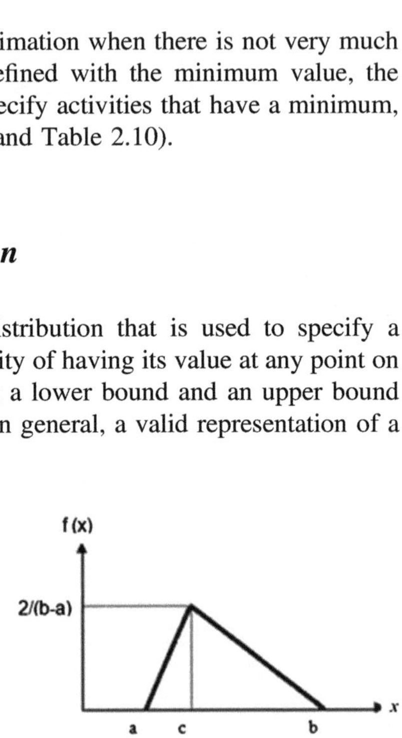

Fig. 2.9 Triangular distribution function

Table 2.10 Triangular distribution function

Triangular	$Trian(a, b, c)$
Probability density	$f(x)\begin{cases} \frac{2(x-a)}{(b-a)(c-a)} & a \leq x \leq c \\ \frac{2(b-x)}{(b-a)(b-c)} & c < x \leq b \\ 0 & o.c \end{cases}$
Cumulative distribution	$F(x)\begin{cases} 0 & x < a \\ \frac{(x-a)^2}{(b-a)(c-a)} & a \leq x \leq c \\ 1 - \frac{(b-x)^2}{(b-a)(b-c)} & c < x \leq b \\ 1 & b < x \end{cases}$
Mean	$\frac{a+b+c}{3}$
Variance	$\frac{a^2+b^2+c^2-ab-ac-bc}{18}$

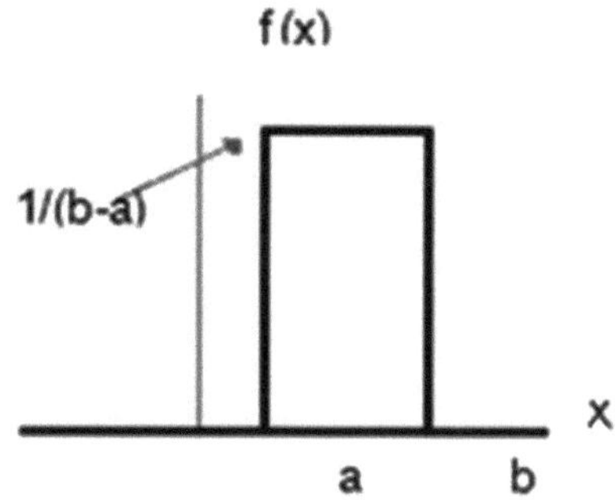

Fig. 2.10 Uniform distribution function

Table 2.11 Uniform distribution function

Uniform	$U(a,b)$
Probability density	$f(x) = \begin{cases} \frac{1}{b-a} & a \leq x \leq b \\ 0 & o.c. \end{cases}$
Cumulative distribution	$F(x) = \begin{cases} 0 & x < a \\ \frac{x-a}{b-a} & a \leq x \leq b \\ 1 & b < x \end{cases}$
Mean	$\frac{(a+b)}{2}$
Variance	$\frac{(b-a)^2}{12}$

random phenomenon. It is used when the distribution is unknown and there is only information about the extreme values (Fig. 2.10 and Table 2.11).

2.6.7 Weibull Distribution Function

The Weibull distribution function is a family of distributions that depend on two parameters: the shape parameter α and the scale parameter β. When the shape parameter $\alpha = 1$, both the Weibull distribution and the Gamma distribution are reduced to the negative exponential distribution. This is used for modeling process times and also for modeling the reliability of an item of equipment by defining the time that elapses until the equipment breaks down. An additional parameter can be introduced by replacing the Weibull random variable X by $X-a$, where a is a location parameter that represents a threshold or guarantee time (Table 2.12).

As you can see in Fig. 2.11, this distribution has different shapes depending on the value of the scale parameter β, although the shape parameter α can also be made to vary and likewise obtain different shapes.

In numerous situations, the empirical probability distribution has such a shape that there is not a standard distribution that properly represents the behavior of the process. The options that can be posed for its formalization are various:

Table 2.12 Weibull distribution function

Weibull	*Weibull* (α, β)
Probability density	$f(x) = \begin{cases} \alpha\beta^{-\alpha}x^{\alpha-1}e^{-(x/\beta)^{\alpha}} & x \geq 0 \\ 0 & x < 0 \end{cases}$
Cumulative distribution	$F(x) = \begin{cases} 1 - e^{-(x/\beta)^{\alpha}} & x \geq 0 \\ 0 & x < 0 \end{cases}$
Mean	$\frac{\beta}{\alpha}\Gamma\left(\frac{1}{\alpha}\right)$
Variance	$\frac{\beta^2}{\alpha}\left\{2\Gamma\left(\frac{2}{a}\right) - \frac{1}{\alpha}\left[\Gamma\left(\frac{1}{\alpha}\right)\right]^2\right\}$

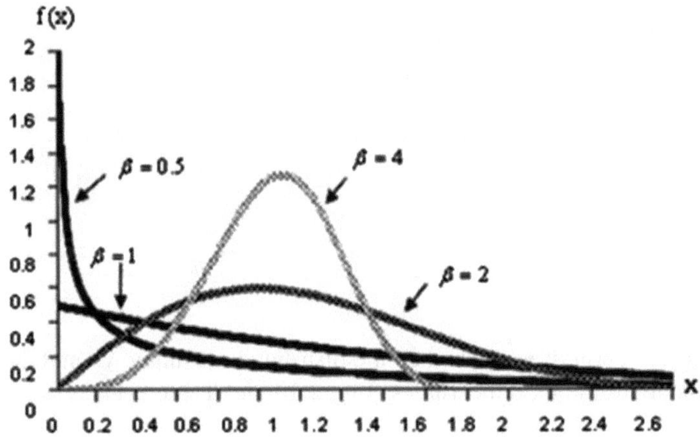

Fig. 2.11 Weibull distribution function ($\alpha = 1/2$)

- Directly employing the empirical probability distribution with the advantages and drawbacks already described for this option.
- Rejecting the values of the sample that are clearly atypical: This is possible, always provided the loss of information can be assumed in the context of the process that is being modeled.

If the histogram has several dominant areas, one can try to separate and adjust it in several cases. In other words, a different distribution shall be adjusted in each one of the dominant areas (Law 2006; Altiok and Melamed 2007) obtaining a multi-modal distribution. If p_j is the proportion of samples in each dominant area and fj (x) the probability density function in each one of the areas, the overall probability density function shall be

$$f(x) = \sum_{j=1}^{n} p_j f_j(x)$$

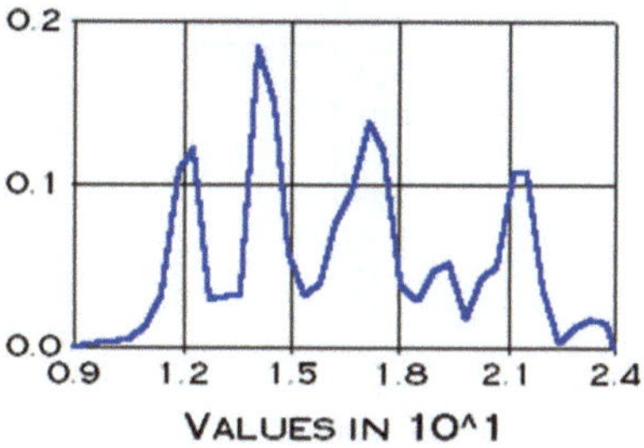

Fig. 2.12 Histogram of the weight in kg/mm of steel coils provided by a supplier (reproduced with the kind permission of Siderúrgica del Mediterráneo S.A.)

Table 2.13 Set of adjusted pdfs

Range (kg/mm)	% of samples	Adjusted pdf
[9–13.5]	18.56	Normal, $\sigma = 0.64302$ $\mu = 12.193$
[13.5–15.5]	24.03	Log-normal, $\sigma = 0.02787$ $\mu = 2.6743$
[15.5–18.5]	28.78	Triangular, m = 17.51 a = 15.358 b = 18.626
[18.5–24]	28.61	Normal, $\sigma = 1.2976$ $\mu = 21.095$

Example 2.2 Modeling of the weight of steel coils
A lot of steel works, especially ones that produce steel for the car industry, work with coils, in other words, their raw materials can be unprocessed steel coils. In the factory there are semi-processed coils and the end product that is delivered to the customer, i.e. processed coils (end product).One of the most important aspects that must be borne in mind when planning production operations is the definition of the physical characteristics of the coil: width, length, thickness, weight. Figure 2.12 is the histogram for a sample of the weight in kilograms per millimeter of width of 1200 steel coils. From this value, the weight of the coil can be obtained by multiplying the diameter of the coil and its length by its width, if its thickness is known.

It is not possible to obtain a unimodal probability density function that fits the histogram. Accordingly, the option was taken to obtain a different adjustment for each one of the four dominant areas. The final result was (Table 2.13).

Figure 2.13 shows the adjustment obtained for one of the dominant areas.

2.7 Discrete Distribution Functions

2.7.1 Bernoulli Distribution Function

Bernoulli distribution is applied in cases where there are two possible states. The probability of one state is p and that of another state $q = 1 - p$. The phenomena that define them are, among others:

- Whether or not the piece that exits the process is defective.
- Whether or not an employee comes to work.

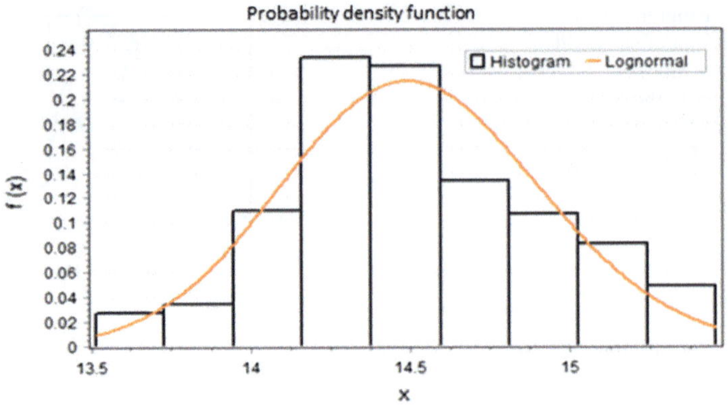

Fig. 2.13 Histogram and pdf adjusted for the second dominant area (reproduced with the kind permission of Siderúrgica del Mediterráneo S.A.)

- Whether or not an operation requires a secondary process, a reoperation (Fig. 2.14 and Table 2.14).

2.7.2 Discrete Uniform Distribution Function

This is used when all the values in the [*i*, *j*] range have an equal probability. It is employed as a first model, when we only have information about the limits of the range (Table 2.15).

2.7.3 Binomial Distribution Function

Binomial distribution is a discrete distribution that expresses the result of n separate experiments. It is essentially the sum of n Bernoulli experiments. Let us suppose

Fig. 2.14 Bernoulli probability function (p = 0.6)

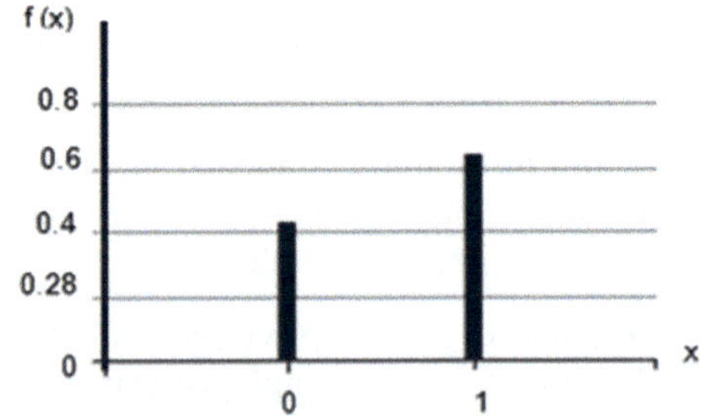

Table 2.14 Bernoulli distribution function

Bernoulli	*Bernoulli (p)*
Probability function	$f(x) = \begin{cases} 1-p & if \quad x=0 \\ p & if \quad x=1 \\ 0 & o.c \end{cases}$
Cumulative distribution	$F(x) = \begin{cases} 0 & if \quad x<0 \\ \sum_{i=0}^{[x]} \binom{n}{i} p^i (1-p)^{n-i} & if \quad i \leq x \leq j \\ 1 & j<x \end{cases}$
Mean	p
Variance	$p(1-p)$

Table 2.15 Discrete uniform distribution function

Discrete uniform	$UD(i,j)$
Probability function	$f(x) = \begin{cases} \frac{1}{j-i+1} & x \in \{i, i+1, \ldots, j \\ 0 & o.c \end{cases}$
Cumulative distribution	$F(x) = \begin{cases} 0 & if \quad x<i \\ \frac{[x]-i+1}{j-i+1} & if \quad i \leq x \leq j \\ 1 & j<x \end{cases}$
Mean	$\frac{(i+j)}{2}$
Variance	$\frac{(j-i+1)^2-1}{12}$

that an experiment that has two possible results is done n times ($n > 0$). Also, let us suppose that the probability of obtaining a particular result, (let us call it result ***a***) for any experiment is p, and the probability of the other result is $q = 1 - p$ (let us call it result b).

Therefore, result a may appear a number of times between 0 and n, as can result b. Binomial distribution specifies the probability of result a occurring k times. Some phenomena that can be defined using this distribution are:

- The number of defective pieces in a batch.
- The number of customers of a particular type that enter the system (Table 2.16).

2.7.4 *Poisson Distribution Function*

The frequency of the appearance of events in an arrivals process can be formalized by specifying the time between two successive arrivals or the number of arrival events per range.

Table 2.16 Binomial distribution function

Binomial	$Bin(n, p)$
Probability function	$f(x) = \begin{cases} \binom{n}{x} p^x (1-p)^{n-x} & x \in \{0, 1, \ldots, n\} \\ 0 & o.c \end{cases}$
Cumulative distribution	$F(x) = \begin{cases} 0 & if \quad x<0 \\ \sum_{i=0}^{[x]} \binom{n}{i} p^i (1-p)^{n-i} & if \quad i \leq x \leq j \\ 1 & \quad j<x \end{cases}$ Where $[x]$ indicates the largest integer $\leq x$
Mean	np
Variance	$np(1-p)$

- *Time between 2 successive arrival events:* in general, the time between two consecutive independent arrival events usually responds to an exponential distribution.
- *Number of arrival events per range:* instead of describing the time between arrival events, the number of events in a range of constant time is described. Note, for example, that it is not possible to describe by means of an exponential distribution the arrival of material at a production unit when it is transported on *pallets* with a number of variable pieces, as the time between the arrival of one piece and the next one is 0. Poisson distribution is one of the most used to describe this type of behavior. This distribution was originally developed for modeling the phone calls of a telephone exchange. Other phenomena that can be modeled are:

1. The number of temporary entities that arrive per unit of time.
2. The total number of defects in a piece.
3. The number of times that a resource is interrupted per unit of time (Table 2.17).

2.7.5 Geometric Distribution Function

Geometric distribution describes the number of experiments with p probability of success, which must be carried out until a particular result is obtained. Some examples of phenomena that can be modeled with this distribution are:

- The number of machine cycles until it breaks down
- The number of pieces inspected until one is found with defects
- The number of customers served until one of a particular type is found (Table 2.18).

Table 2.17 Poisson distribution function

Poisson	$Poisson(\lambda)$
Probability function	$f(x) = \begin{cases} \frac{e^{-\lambda}\lambda^x}{x!} & x \in \{0, 1, \ldots\} \\ o & o.c. \end{cases}$
Cumulative distribution	$F(x) = \begin{cases} 0 & x<0 \\ e^{-\lambda}\sum_{i=0}^{[x]}\frac{\lambda^i}{i!} & o \leq x \end{cases}$
Mean	λ
Variance	λ

Table 2.18 Geometric distribution function

Geometric	$Geom(p)$
Probability function	$f(x) = \begin{cases} p(1-p)^x & x \in \{0, 1, \ldots\} \\ 0 & o.c \end{cases}$
Cumulative distribution	
Mean	$\frac{(1-p)}{p}$
Variance	$\frac{(1-p)}{p^2}$

2.8 Development of a Statistical Model

The development of a statistical model is very important for the design of a simulation model. In the case of discrete event simulation there are items such as the arrivals of customers or elements of interest into a system (that is going to be simulated), the times between the arrivals of the customers or elements into the system, time in the system, service time etc. These concepts shall be covered in more detail in later chapters.

The steps that a statistical model contains are:

1. Collection and analysis of the data
2. Adjustment of a distribution function
3. Validation of the adjustment

Example 2.3

1. *Data collection*

In modeling the random part, only the data referring to the process have to be recorded, without considering either the causes of the random activity or its effect. The time between arrivals at a tollbooth is presented below (Table 2.19).

2. *Data analysis*

IID: in simulation we assume that the values of the data samples are IID: Independent Identically Distributed Values, which means:

Independent: the set of values is not correlated

Identically distributed: follow the same probability distribution (Fig. 2.15).

Table 2.19 Time between arrivals MM1 in sec*100

0.50	3.35	20.85	7.81	0.44	0.03	3.82	7.09	3.02	2.80
2.08	6.53	52.53	10.23	0.76	0.00	28.21	15.51	4.86	10.41
5.25	11.67	46.23	28.06	6.05	4.82	46.36	2.90	5.47	17.42
7.20	41.15	9.54	4.88	19.10	9.17	0.83	7.43	9.98	4.11
10.28	23.44	6.19	2.39	7.57	12.97	12.62	7.65	18.49	6.95
1.08	9.89	5.49	2.16	14.18	11.89	12.73	0.51	14.61	27.01
1.91	18.77	4.98	6.41	1.45	1.71	5.21	2.89	8.38	3.50
2.86	17.60	4.89	11.74	15.31	36.64	3.62	21.78	2.15	6.70
17.13	0.11	17.58	1.30	2.44	9.59	1.74	5.02	6.46	18.76
1.49	7.92	4.03	3.13	1.67	23.31	3.13	9.35	0.10	0.51

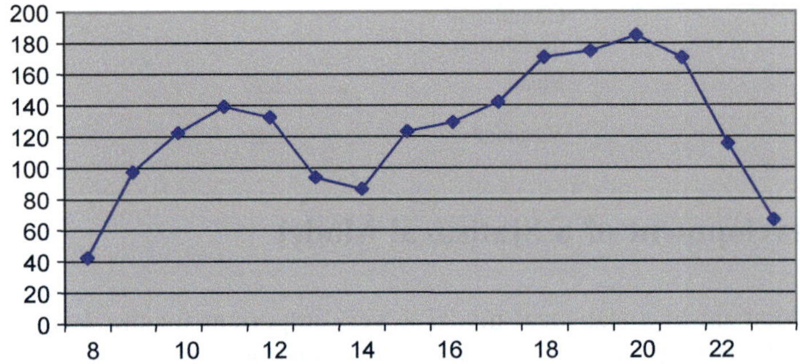

Fig. 2.15 Number of arrivals per time segment

3. *Validation of the adjustment*

To be able to validate the adjustment of the data, goodness of fit tests are used such as the chi square, Kolmogorov Smirnov or the Anderson Darling test. At the present time, one or more than these tests are built into popular simulation packages, such as Promodel, which has Stat Fit that, as mentioned in said software, has the following use:

- Curve fitting. It helps you to find the best distribution to represent the data. **Stat::fit** uses the most commonly known goodness of fit tests such as:

  a. Anderson-Darling
  b. Chi-Square
  c. Kolmogorov-Smirnov

- Determining the number of replications to run a simulation model.
- Determining the size of the sample for taking process and transportation times.

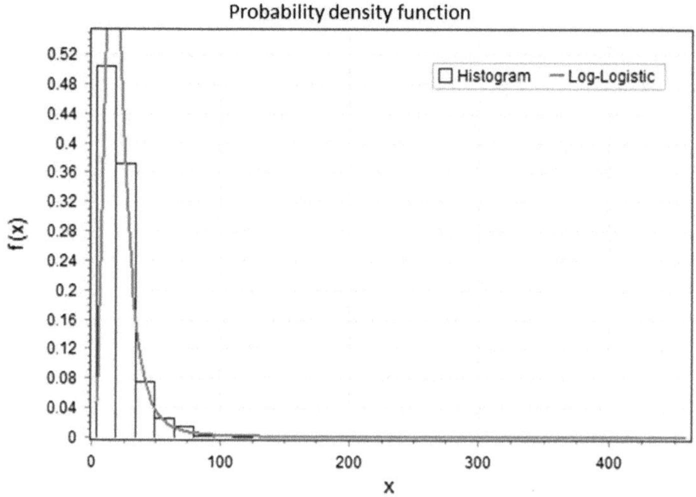

Fig. 2.16 Histogram and pdf LogLogistic

- Graphing the input data, graphing all the probability distributions that can be used, drawing up descriptive statistics for the data.
- An excellent option to disseminate statistical thought.[1]

In the case of SIMIO, this is left to the user's criteria. It does not have built-in data adjustment software.

Example 2.4 Modeling of the time between arrivals at a port terminal
Figure 2.16 shows the histogram and the adjusted probability density function (LogLogistic, $\alpha = 3.4$ $\beta = 19.23$) for a sample of 1687 values for times between arrivals in hours of boats in a port terminal of the Port of Barcelona (Spain).

A total of 8 values from the sample have a time between arrivals of more than 110 h. If the objective of the study is to analyze the behavior of the terminal in periods of normal or high workload, the elimination of these 8 values in the sample (0.5% of the total) will not have a significant impact on the results of the study, given that they correspond to periods of little activity. Figure 2.17 shows the histogram for the real sample without the above 8 values and the adjusted probability density function (LogLogistic, $\alpha = 3.6$ $\beta = 19.01$).

[1]http://www.promodel.com.mx/statfit.php.

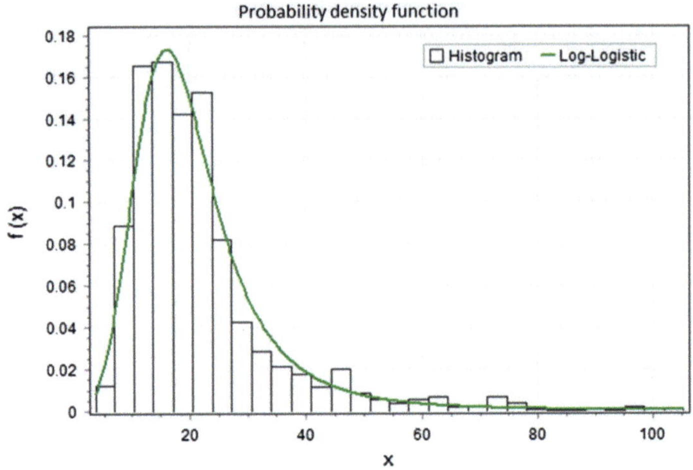

Fig. 2.17 Histogram of the real sample

2.9 Statistical Analysis of the Simulation Results

Once the simulation has been done it is important for a statistical analysis of the results of the simulation to be carried out. In many simulation studies too much time and money is spent on the development of the model and the programming and very little effort is made to properly analyze the results. In fact, a very common mode of operation is to make a single simulation run of a somewhat arbitrary length and then use the results of said run as true, however, owing to the randomness of the variables there can be large variations in the results that entail wrong inferences about the real problem.

That is why it is necessary to develop one or more experiments and one or more replications in each one of them. We can see this pattern in schematic form Fig. 2.18.

Figure 2.18 illustrates the difference between an experiment and a replication. As can be appreciated, the replications are in the same experiment. When carrying out a simulation study, a series of parameters are considered, for example, the time between arrivals. Along these lines, a change is made in the time between arrivals or a variation can also be considered in the number of servers. The implementation of each one of these changes corresponds to the development of different experiments to evaluate different scenarios. The replications are run from the same model without making any changes to the parameters: These replications will give different results owing to the use of different series of random numbers. Every replication produces statistical results that differ from those produced by other replications, and said results can be analyzed throughout the entire set of replications.

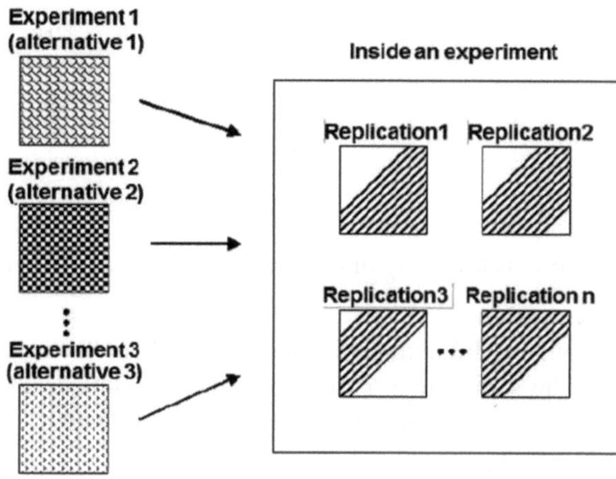

Fig. 2.18 Experiments and replications in a simulation model. *Source* Law 2006

In this part of the experiment we are interested in being able to answer the following questions:

1. How many experiments need to be carried out?
2. How many replications are needed for the simulation?

The random nature of the simulation entails a stochastic process in the results that we can call $Y_1, Y_2,\ldots Y_m$ for a simple run. For example, Y_i could be a delay in the arrival of the i-th job in the simple queuing system. C_i is also defined as the cost of operating an inventory in the i-th month. Y_{ij} are random variables qj that are not, in general, independent or identically distributed (IID). Thus, many of the formulas of classic statistics shall not be directly applicable to the analysis of the simulation's output data.

Example 2.5 (Law 2010)
A queuing system is considered where Y_1 is the delay of customer 1, Y_2 is the delay of customer 2, etc. In this system the delays in the queue shall not be independent, as a long delay for a customer waiting in a queue will tend to even further delay the next waiting customer. Assuming that the simulation starts at zero time without any customers in the system, as usually done.Then, delays in the queue at the start of simulation will tend to be shorter than delays at the end, hence the delays are not identically distributed.

Let $y_{11}, y_{12},\ldots, y_{1m}$ be results from running the simulation for random variables $Y_1, Y_2,\ldots Y_m$ with the specific random numbers $u_{11}, u_{12}, \ldots$.If the simulation is run with a different set of random numbers $u_{21}, u_{22}, \ldots$, then a set of different results $y_{21}, y_{22},\ldots, y_{2m}$ is obtained from the same random variables. $Y1, Y2,\ldots Y_m$.

The two sets of results are not equal as different random numbers were used in two runs and, accordingly, two different samples were produced from the same probability distributions.

In general, assuming that n independent replications or simulation runs each one of size m, that would, in the example, mean simulating the delays of m customers, obtaining the following observations:

$$
\begin{array}{c}
y_{11}, \ldots, y_{1i}, \ldots, y_{1m} \\
y_{21}, \ldots, y_{2i}, \ldots, y_{2m} \\
\ldots\ldots\ldots\ldots\ldots\ldots \\
y_{j1}, \ldots, y_{ji}, \ldots, y_{jm} \\
\ldots\ldots\ldots\ldots\ldots\ldots \\
y_{n1}, \ldots, y_{ni}, \ldots, y_{nm}
\end{array}
$$

Where y_{ji} is the delay of customer i in the replication j for i = 1,…, m and j = 1,…, n

Observation 1
Although a different set of random numbers is used in every replication, everyone uses the same initial conditions, and the statistical counters for the simulation is restarted every time a new replication is initiated.

Observation 2
The observations of a replication in particular (line) are clearly not IID. However, note that $y1i$, $y2i$,.., yni of the i-th column, are IID observations of the random variable Yi, for i = 1,2, …, m.

In general terms, we can assert that the replications are independent of each other, and that the observations of each replication have the same joint distribution.

This independence throughout the runs is the key to simplifying the data analysis of the simulation. Accordingly, in general terms, the purpose of analyzing the data from the simulation results is to use the observations yji (i = 1,2,.., m; j = 1,2,.., n) to infer the characteristics of the random variables $Y1$, $Y2$,…Ym.

There can be two types of simulation, terminating or non-terminating, and in each case some of the problems involved must be solved for each situation. As Currie and Cheng (2013) mention, "Examples of terminating simulations are the end of a working day or the occurrence of some random event. The time of the end event need not be deterministic and consequently the length of the output data is not necessarily the same for each run of the simulation. Conversely, non-terminating simulations have no defined end event and in such situations we are usually interested in the steady-state behavior of the system. The difficulty with analyzing non-terminating simulations is determining when the steady-state has been reached."

Examples of terminating and non-terminating simulations:	
Simple queuing system	Non-terminating
Outbreak of an A(H1N1) epidemic	Terminating
Bank branch, customer care	Terminating
Blood bank	Non-terminating
Supply chain	Terminating

2.10 Conclusions

This chapter shows some of the basic and essential concepts of statistics in order to develop a simulation model with everything it requires, initial data, data adjustment, runs and experiments, model verification and validation, as well as the interpretation of the results.

As simulation is a stochastic process there is always something more we could say about it. If the reader would like to further information, we recommend they consult the following references.

References

Altiok, T., & Melamed, B. (2007). *Simulation modeling and analysis with ARENA*. New York: Academic Press.

Banks, J. (ed.). (1998). *Handbook of simulation*. New York: Wiley.

Barton, R. (2004). Designing simulation experiments. In *Proceedings of the 2004 winter simulation conference* (pp. 73–79).

Carson, J., & Banks, J. (1993). Discrete- event system simulation. Englewood Cliffs: Prentice Hall.

Coss, B. R. (2003). *Simulación*. Limusa: Un enfoque práctico.

Currie, C., & Cheng, R. (2013) A practical introduction to analysis of simulation output data. In R. Pasupathy, S.-H. Kim, A. Tolk, R. Hill & M. E. Kuhl (Eds.), *Proceedings of the 2013 winter simulation conference* (pp. 328–341).

Flores, I., & Elizondo, M. (2007). *Apuntes de simulación*, Facultad de Ingeniería UNAM.

González, M. C. (1996). *Modelos y simulación*. UNAM: ENEP Acatlán.

Gordon, G. (1991). *Simulación de sistemas*, 6ta. reimpresión de la 1ª. Edición, Diana.

Guasch, A., Piera, M. A., Casanovas, J., & Figueras, Y. J. (2003). *Modelado y simulación: aplicación a procesos logísticos de fabricación y servicios*. 2a. ed., Barcelona, Ediciones UPC.

Kelton, D., & Barton, R. (2003). Experimental design for simulation. In *Proceedings of the 2003 winter simulation conference* (pp. 59–65).

Kolmogorov, A. N. Y., & Uspenskii, V.A. (1987). *Algorithms and randomness*, Ed. Teor. Veroyatnost. i Primenen.

Law, A. (2003). How to conduct a successful simulation study. In *Proceedings of the 2003 winter simulation conference* (pp. 66–70).

Law, A. (2004). Statistical analysis of simulation output data: The practical state of the art. In *Proceedings of the 2004 winter simulation conference* (pp. 67–72).

Law, A. (2006). *Simulation modeling and analysis with expertfit software*. New York: Mc. Graw-Hill.

Law, A. (2010). Statistical analysis of simulation output data: The practical state of the art. In *Proceedings of the 2010 winter simulation conference* (pp. 65–73).

Law, A., & Kelton, D. (2000). *Simulation modelling and analysis*. New York: Mc. Graw-Hill.

Lécuyer, P. (1990). Random numbers for simulation. *Communications of the ACM, 33*, Núm. 10, 85–97.

Lehmer, D. H. (1951). Mathematical methods in large-scale computing units. In *Proceedings of a second symposium on large-scale digital calculating machinery* (pp. 141–146). Cambridge: Harvard University Press.

Robinson, S., & Bhatia, V (1995). Secrets of successful simulation projects. In *Conference: Simulation conference proceedings, Winter WSC '95* Proceedings of the 27th conference on Winter simulation, pp. 61–67

Skoogh, A., & Johansson, B. (2008). A methodology for input data management in discrete event simulation projects. In S. J. Mason, R. R. Hill, L. Mönch, O. Rose, T. Jefferson & J. W. Fowler (Eds.), *Proceedings of the 2008 winter simulation conference* (pp. 1727–1735).

Taha, H. (2004). *Investigación de Operaciones*, 7ª. ed., Prentice-Hall.

Trybula, W. J. (1994). Building simulation models without data. *Proceedings of the IEEE International Conference of Systems, Man and Cybernetics,* 209–214.

Chapter 3
Modeling Discrete Event Systems Using Petri Nets

M. Narciso and M.A. Piera

3.1 Introduction

The main characteristic of the behavior of a Discrete Event System (DES) is that the System's state variables do not change value until an event happens, in other words, the event generates an instant change in the state of the system. Time advances from event to event and events happen chronologically but not necessarily at regular time intervals. The complexity in these types of systems lies in the fact that any decision can block, freeze, delay, enable/disable future events.

Consider, by way of example, the process of airplane arrivals and departures at an airport that only has one runway. It is easy to see how the runway will be blocked until an airplane has finished landing and how this can delay scheduled take-offs with airplanes being forced to wait at the head of runway. In this same example we observe that we cannot accurately predict how the state of the system will evolve (simply think of the evolution over time of the numbers of airplanes waiting at the head of the runway), as the time required for runway use depends on conditions that are hard to represent, such as how much experience the pilot has, wind conditions, the exact weight of the airplane at the instant of take off, among other things, and that can have an influence on deviations from the nominal expected time for the take-off operation. In addition to these uncertainties, which can be modeled as stochastic activities that influence the time an event may last (start or end of an activity), there are also physical and logical constraints that may delay the occurrence of an event, known as concurrence, synchronization and parallelism, for a variety of activities that have an effect on the system's evolution.

In this sense, the stochastic, dynamic and asynchronous nature of discrete event systems demands a modeling formalism that considers all the aforementioned characteristics that influence the system's evolution and allows us to represent both its structure and its behavior as a function of the possible settings. This should all favor the maintenance of the model, eliminating or adding new events in

I.F. De La Mota et al., *Robust Modelling and Simulation*,
DOI 10.1007/978-3-319-53321-6_3

accordance with any changes to the system, or to the operational context, that could ultimately have an influence on said model.

Therefore, one requirement of the DES modeling formalism is that it must allow us to represent all the events that influence the system and the causal relationships between them so as to represent its entire behavior.

3.2 Petri Nets

Petri nets (PN) are a modeling formalism that enables us to naturally represent a DES. Models of discrete event systems are essentially based on the concepts of events and activities. An event corresponds to a change in the value of the system's state variables and an activity encapsulates what happens between two events. Although a PN is not the only formalism that will enable us to represent events and activities, it facilitates our formal representation of parallelism and synchronization (Silva and Valette 1989; Zaremba and Prasad 1994; Zimmerman 1995; Zimmermann et al. 1996; Zhou and Venkatesh 1999; Petri Nets World 2015).

Other aspects that contribute to potentiating PNs in both the modeling and the quantitative analysis of DES, as well as for the verification, validation and analysis of the simulation results, are (Guasch et al. 2003):

- They allow us to study structural aspects of the system, such as jammed situations or the achievability of certain states.
- They make it possible to immediately determine all those events that might occur when the system is in a certain state and all the events that could be unleashed by the occurrence of a particular event.
- They allow us to formalize a system at different levels of abstraction, in accord with the modeling objectives.
- They enable the description of a complex system using the bottom-up methodology: development of the complete system model based on the PN (submodels) of the subsystems that have already been developed and verified.
- They constitute a graphic modeling formalism with very few syntactic rules.
- They enable us to find the possible paths to achieving a final state starting from an initial state, and to know the cost of each one of the paths.
- They make it possible to obtain the set of possible states that can be achieved starting from an initial state.

3.2.1 *Description of Petri Nets*

A PN is a particular case of the directed, weighted and bipartite graph that uses the following elements of representation:

- **Place nodes:** are graphically represented by circles or ellipses and can be used both to describe a system's queues (warehouses, *buffers*, *stocks*, etc.) and to describe conditions on the state in which the elements or resources that make up the system are to be found. Figure 3.1 gives the graphic representation of a PN with 5 place nodes, known as P1, P2, P3, P4 and P5.
- **Transition nodes:** are represented by rectangles and can be used for modeling the events that appear in the dynamics of a system. The PN of Fig. 3.1 contains a single transition node known as T1.
- **Directed arcs:** are represented by arrows and make it possible to connect a place node with a transition node, or a transition node with a place node, but never two nodes of the same type. The PN in Fig. 3.1 contains five directed arcs, three of which connect place nodes P1, P2 and P3 to transition T1, and two of which connect transition node T1 to place nodes P4 and P5.
- **Weights**[1]: the arcs that connect the place nodes with the transition nodes usually have an associated weight, which allows us to describe, for example, the conditions required for the event represented by the transition node to be able to occur. Similarly, the arcs that connect the transition nodes with the place nodes usually have an associated weight, which makes it possible to describe the changes to the state of the system as a consequence of the occurrence of the event represented by the transition node. The arcs represented in the PN of Fig. 3.1 have weights of value 1, 2 and 3.
- **Tokens**[2]: are graphically represented as points inside a place node and allow us to model, for example, the number of elements (pieces, resources, personal banking, etc.) in a place node, or the state of a condition (true or false) that indicates the fulfillment or not of said state of a condition. In the PN in Fig. 3.1, place node P1 has 4 tokens, place nodes P2 and P4 have 1 token each, place node P3 has 3 tokens and place node P5 has 2 tokens.
- **Marking:** a marking represents any arbitrary distribution of tokens in the place nodes. The initial distribution of tokens in the place nodes is called initial marking. In the specific case of the PN in Fig. 3.1, the marking can be represented as a vector with the following values [4, 1, 3, 1, 2].

In a PN, we say that a node X is an input node from another node Y, if and only if there is an arc directed from X to Y. Similarly, a node X is said to be an output node from another node Y, if and only if there is an arc directed from Y to X. Thus, we can be speak of input place nodes, output place nodes, input transitions, output transitions, input arcs and output arcs.

Considering again the PN in Fig. 3.1:

- Place nodes P1, P2 and P3 are input nodes for transition T1.
- Place nodes P4 and P5 are output nodes of transition T1.

[1]When the arc weight is missing, by default it's value is considered as 1.

[2]Also called *tokens* in the original nomenclature.

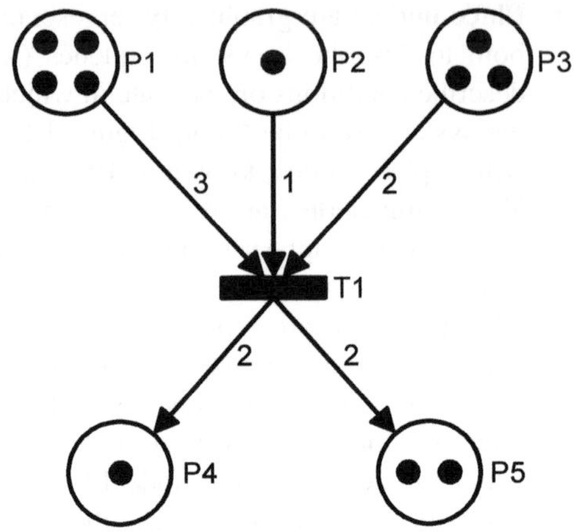

Fig. 3.1 Graphic representation of a PN

- The arcs that go from place nodes P1, P2 and P3 to transition T1 are input arcs to the transition and have a weight of 3, 1 and 2 respectively.
- The arcs that go from transition T1 to place nodes P4 and P5 are output arcs of the transition and have a weight of 2 each.

Although place nodes and transition nodes can be interpreted in different ways (Wang 1998), in this book the events shall be represented by transitions, the place nodes shall enable us to represent the state of the system, and the activities[3] shall be represented by place nodes encapsulated between 2 transitions.

3.2.2 Formal Definition of Petri Nets

In mathematical terms, a PN is defined as a tuple composed of five elements (Proth and Xie 1996; Wang 1998; Guasch et al. 2003):

$$\mathrm{RdP} = (\mathrm{P}, \mathrm{T}, \mathrm{A}, \mathrm{W}, \mathrm{M0})$$

where:

P = {P1, P2, P3, …, Pnp}: Non-empty finite set with place nodes (*np* is the number of place nodes in the PN).

T = {T1, T2, T3, …, Tnt}: Non-empty finite set with transition nodes, (*nt* is the number of transition nodes in the PN).

[3]According to the abstraction made of the system to be modeled, an activity can be represented in a PN as a place node or a transition node.

A = {a1, a2, a3, …, ana}: Non-empty finite set with arcs, (*na* is the number of arcs in the PN).
A is a subset of the Cartesian product of sets *P* and *T*:

$$A \subset (P \times T) \cup (T \times P)$$

where the first element of the ordered pair corresponds to the origin node and the second element to the destination node. The two nodes have to be different types, accordingly if a node is a transition type the other must be a place type and vice versa.

W: Ai → {1, 2, 3, …}: Weight associated with each arc Ai ∀ i = 1, 2, …, na.

M0 = [p1, p2, p3, …, pnp]: Initial marking, where pj is the number of tokens in the *j*th place node Pj: M(Pj) = pj ∀ j = 1, 2, …, np.
The state of a system, after the occurrence of an event, is totally determined by the number of tokens in each place node, and can be mathematically described by the vector:

$$\mathrm{Ms} = [\mathrm{p1}, \mathrm{p2}, \mathrm{p3}, \ldots, \mathrm{pnp}] \quad \forall\, \mathrm{s} = 1, 2, \ldots, \mathrm{nm}$$

where *nm* represents the total number of markings (states) of the system.
Thus, in the example of Fig. 3.1 the initial state can be represented as follows:

$$\mathrm{M0} = [4, 1, 3, 1, 2]$$

3.2.3 Behavior or Dynamics of Petri Nets

A PN can be treated as a games board where the tokens represent counters (which can only be put on the place nodes) (Jensen 1997). Each transition represents a potential movement in the game. A movement is possible if and only if each input place node of the transition contains at least the number of tokens prescribed by the weight of the corresponding input arc.

The rules for simulating the behavior or dynamics of a PN are (Proth and Xie 1996; Guasch et al., 2003):

- A transition It is enabled if each one of the place nodes Pj connected to the input contains at least W(Pj, It) tokens. Where each W(Pj, It) represents the weight of the arc that joins node Pj to transition It. If its su does not appear on an arc, it is taken as being 1.

- An enabled transition can be fired at any instant in time.
- As a result of firing an enabled transition, W(Pj, It) tokens are eliminated from each node Pj in the It input, and W(It, Pk) tokens are added to each Pk node from the It output. Where W(It, Pk) corresponds to the weight of the arc that joins the It transition to the Pk node.

It is then said that a transition is *disabled* when there are less tokens in any of the input place nodes of the transition, for which the respective weights of input arcs to said transition prescribe, otherwise it is said that is *enabled*.

When a transition is enabled, the corresponding movement can take place. If this happens, we say that the transition has been fired. The effect of the occurrence of a transition is that the tokens are eliminated from the input place nodes and are added to the output place nodes. The number of tokens eliminate is specified by the weight of the corresponding input/output arc. It is important to point out that there is no relationship between the tokens eliminated from the input place nodes and those added to the output place nodes. The total number of tokens eliminated from the various input place nodes could be different from the number of tokens added to the various place output nodes.

When there are one or more enabled transitions for a given marking Ms, it means that every one of these transitions can be fired. Moreover, if there are enough tokens in the input place nodes for each one of them, so that each one of the transitions can obtain its own tokens without having to share them with the other transitions, it is said that the transitions are *concurrently enabled* in marking Ms. This means that the transitions can be fired "at the same time" or "in parallel".

It is worth mentioning that two transitions are only concurrently enabled if they are independent, in the sense that they can operate on disjoint sets of tokens. In this case the effect, when all the concurrently enabled transitions are fired, is that the resulting immediate actions are simultaneous and this effect is symbolically obtained as the "sum" of the effects of the individual transitions. The order in which each one of the individual transitions occurs does not affect the overall state of the system.

Using the PN from Fig. 3.1 as an example, we can illustrate the new state achieved by firing transition T1, according to the rules of behavior (Fig. 3.2):

The left side of Fig. 3.2 shows an enabled transition:

- The input nodes of the transition contain at least as many tokens as the weight of the arcs that connect them to the transition: P1 contains more than 3 tokens (W (P1, T1) = 3), P2 contains at least 1 token (W(P2, T1) = 1), and P3 contains more than 2 tokens (W(P3, T1) = 2).

The right side of Fig. 3.2 represents the state of the same PN once the transition is fired:

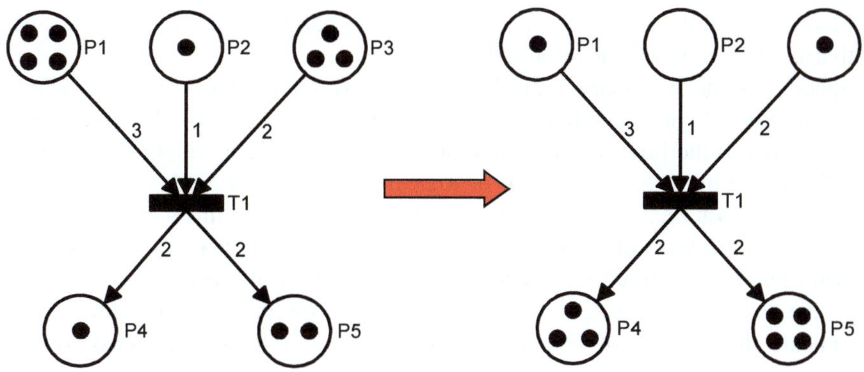

Fig. 3.2 Behavior or dynamics of a PN

- 3 tokens have been eliminated from place node P1 (W(P1, T1) = 3), one token from node P2 (W(P2, T1) = 1), and 2 tokens from place node P3 (W(P3, T1) = 2) 2 tokens have been added to place nodes P4 and P5 (W(T1, P4) = 2 and W(T1, P5) = 2).

The mathematical formalization of the PN allows us to analyze the dynamics of the modeled system based on the observation of possible events that could happen as a result of present state of the system and events that could happen because of the occurrence of a particular event.

Thus, in the example given in Fig. 3.2 we can observe the following states:

$$M0 = [4, 1, 3, 1, 2]$$
$$M1 = [1, 0, 1, 3, 4]$$

where M0 represents the initial state of the system and M1 the state of the system after the firing of transition T1.

3.3 Development of Petri Net Models of Systems

In this section we shall illustrate the concepts of ease of maintenance of models, as well as the relations of concurrency, synchronization and parallelism between the activities that are to be modeled. To this end, we propose the development of a set of submodels for the different functional specifications of a production system, which shall later be integrated to represent the entire flexible production system in PN.

There are many approximations for the modeling of systems using PN but in this section we are proposing the following steps:

1. Identifying all the events that could generate a change in the state of the system that is of interest for the simulation study.
2. Specifying the necessary logical *preconditions* for each event to be able to occur. Each logical condition will be formalized by means of a place node connected to the input of the transition, and with a certain weight on the arc to indicate the number of (physical or logical) resources required.
3. Specifying for each event the changes that shall occur to the state variables of the system because of the occurrence of the event. Said changes are known as *post-conditions* and are formalized by means of a set of place nodes connected to the transition output, with a certain weight on the corresponding arc to indicate the number of resources (both physical or logical) affected by said event.

Example 3.1 PN Model of a production unit

Let us consider a drilling machine with a set of pieces to be processed that are stored in stock S1, and a stock S2 with the already-drilled pieces. The machine has automatisms to be supplied with a piece from stock S1, drill it and as soon as said procedure is over, the piece is also deposited by means of automatisms in stock S2 with infinite capacity.

Under these operating conditions, machine M1 has 2 possible states: *free* and *working* and, in consequence, there are just only 2 possible events that make it possible to represent these changes of state:

1. *Start drilling procedure* event (T1): There are 2 logical preconditions for machine M1 to be able to start a drilling procedure:

   i. The machine must be *free*. It cannot drill 2 pieces at the same time, so it is necessary for the machine to be empty without any piece inside. In Fig. 3.8, this precondition is represented by an input arc of unit weight to transition T1 from place node P2 (machine M1 in *free* state).
   ii. Stock S1 must contain at least one piece awaiting drilling. There is no sense in starting a drilling procedure if there are no available pieces. In the Fig. 3.8, this precondition is represented by an input arc of unit weight to transition T1 from place node P1 (stock S1 with pieces to be processed).

As a consequence of the occurrence of a *Start drilling procedure* event, the state of machine M1 shall go to *working*, and one piece is eliminated from stock S1 that shall go on to be inside the automatisms of the machine. In Fig. 3.5, the output arc of transition T1 to place node P3 allows us to represent the change of state to *working*.

In Fig. 3.3 we can see the changes to be expected as a result of the appearance of a *Start drilling procedure* event. At the top the system is showed before the occurrence of the event (left side). Here we can observe 3 pieces that are awaiting drilling in incoming stock S1 and machine M1 *free*, and the right side illustrates the same system after the occurrence of the event, showing 2 pieces in stock S1 and one piece inside machine M1, being drilled.

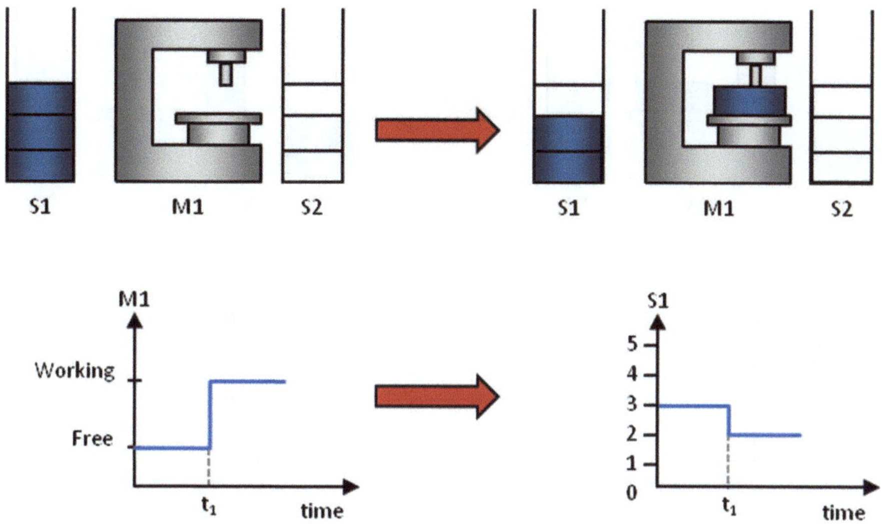

Fig. 3.3 Change of state because of the occurrence of the *start drilling procedure* event (T1)

At the bottom of the same figure, the changes to the state variables when the event happens in a particular instant t_1 have been represented. As can be seen, machine M1 will go from the *free* state to the *working* state, and stock S1 shall see a drop in the number of pieces.

2. *End drilling procedure* Event (T2): There is only one logical precondition for the event to be able to happen, and that is for machine M1 to be in a *working* state. It is not possible for an *End of drilling procedure* event to happen if the machine is in a *free* state. In Fig. 3.5, this precondition is represented by an input arc with a unit weight to transition T2 from place node P3 (machine M1 in *working* state).
3. As a consequence of the occurrence of an *End drilling procedure* event, the state of machine M1 shall go to *free*, and the number of drilled pieces in stock S2 shall be increased by one. In Fig. 3.5, the output arc of transition T2 to place node P2 allows us to represent the change of state to *free*, and the output arc from T2 to place node P4 enables us to represent a new already-drilled piece in stock S2.

Figure 3.4 shows the changes to be expected by the appearance of an *End drilling procedure* event. The top represents the system before the occurrence of the event (left side) where we can observe 2 pieces awaiting drilling in incoming stock S1, machine M1 drilling one piece and the stock of drilled pieces empty. The right side represents the same system after the occurrence of the event. Here we observe 2 pieces in stock S1, machine M1 empty and an already-drilled piece in stock S2. The bottom of the same figure represents the changes to the state variables when the event happens in a particular instant t_2. As can be seen, machine M1 will go from

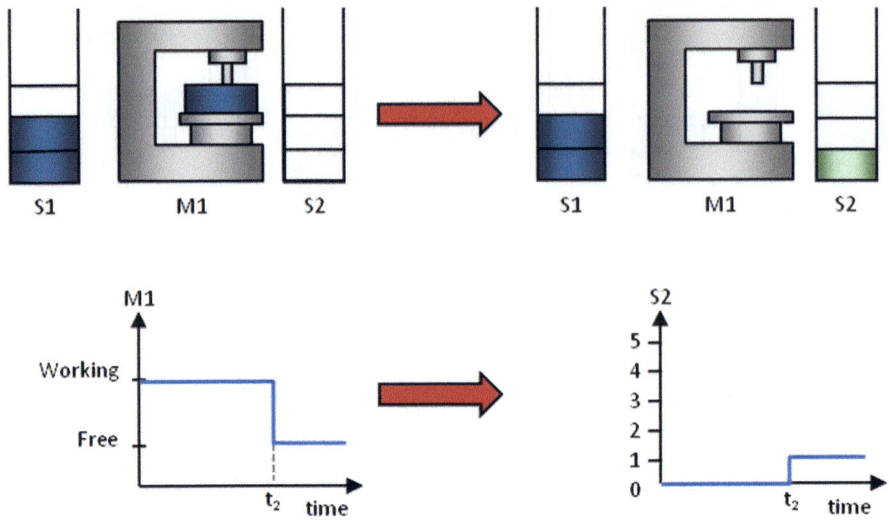

Fig. 3.4 Change of state because of the occurrence of the *end drilling procedure* event (T2)

working state to *free* state, and stock S2 shall see the number of drilled pieces increase by 1.

Figure 3.5 represents the PN for the drilling machine that has been described, while Table 3.1 describes the significance of the place nodes and Table 3.2, the significance of the transition nodes used.

Figure 3.5a shows the PN for the drilling machine considering as an initial condition that there are 3 pieces in stock S1 and that machine M1 is in a *free* state. Under these operating conditions, only transition T1 is found to be activated, as the number of tokens in P1 is higher than the weight of the arc that connects place node P1 to transition node T1 and the number of tokens in P2 is equal to the weight of the arc that connects place node P2 to transition node T1 (M(P1) >= W(P1, T1) and M

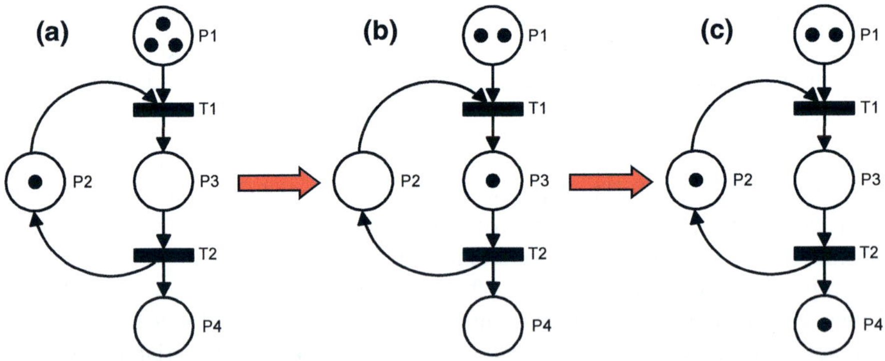

Fig. 3.5 PN model of a production machine

Table 3.1 PN place nodes for drilling machine

Place	Description
P1	Stock S1 of pieces awaiting drilling
P2	Machine M1 in *free* state
P3	Machine M1 in *working* state
P4	Stock S2 of already-drilled pieces

Table 3.2 PN transition nodes for drilling machine

Transition	Description
T1	Start drilling procedure
T2	End drilling procedure

(P2) >= W(P2, T1)). Whereas transition T2 is not activated because in place node P3 there is not at least one token.

In Fig. 3.5b we observe the same PN Model, but after transition T1 has been triggered. We can see that, as a consequence of the transition being triggered, one token has been eliminated in each one of the place nodes connected to the input of transition T1 (in other words, P1 and P2), and one token has been added to the place node connected to the output of the transition (i.e., P3). In (c) we observe the same network model as in (b), but after transition T2 is triggered. We can see that, as a consequence of the transition being triggered, one token has been eliminated from the place node connected to the input of transition T2 (in other words, P3) and one token has been added to each one of the place nodes connected to the output of the transition (i.e., P2 and P4).

Example 3.2 CPN Model for different sequences of production operations on different types of pieces

Consider in this example that 3 different types of pieces have to be processed, each of which requires a sequence of production operations specified in the recipe described in Table 3.3.

The manufacturing system consists of the production operations having to be done on different machines, each one with an incoming and outgoing stock that permits the production activities to be independent between the machines. The stocks leaving the machines correspond to the incoming stock for the machine that must perform the following production operation.

Figure 3.6 shows the elements of the production system where there is a transport subsystem (handling device) that makes it possible to move a processed piece in a machine to the stock for any other machine, or to the output stock S7, if the piece has already gone through all the production operations. Thus, in the case

Table 3.3 Sequence of procedures for type Π_1, Π_2, Π_3 pieces

Type of pieces	Sequence of procedures
Π_1	Adjustment—Pressing—Drilling—Polishing
Π_2	Molding—Milling—Drilling
Π_3	Adjustment—Molding—Milling—Pressing—Polishing

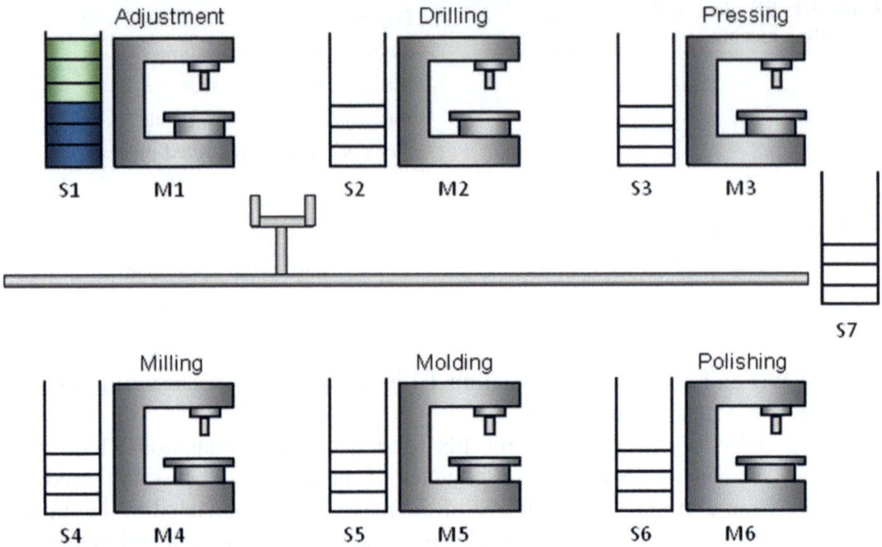

Fig. 3.6 Production system

of type Π_1 pieces, after the adjustment procedure has been performed, the handling device shall transport the piece to stock S3, whereas with type Π_3 pieces, after the adjustment procedure has been done, the handling device shall transport the piece to stock S5. The initial state of the system consists of 3 type Π_1 pieces and 3 type Π_3 pieces in stock S1, that are waiting for the adjustment procedure to be performed on them and 3 type Π_2 pieces that are waiting in stock S5 for a molding procedure to be performed on them.

In Fig. 3.7 we observe the PN that formalizes the 3 sequences of procedures that must be performed for the processing of the pieces stored in stock S1 and S5. Note that stock S1 cannot be represented in PN by a single place node but rather by 2 place nodes (in other words, P1 and P21) in order to be able to differentiate type Π_1 and type Π_3 pieces. Thus, although there is physically a single stock S1, at a logical level 2 stocks are considered, one for type Π_1 pieces (in other words, place node P1) and another stock for type Π_3 pieces (in other words, place node P21).

In the Table 3.4 the meaning of the place nodes is described and in Table 3.5 the meaning of the transition nodes that were used to model the manufacturing system is described.

As can be seen, the states machine *free* (in other words, P2, P5, P8, P11, P25, P28) are common to the different production sequences, as they do not have an assigned type of piece. Each one of these nodes participates in a pattern of behavior of the Decision/Conflict type, where there are two or more transitions that compete to be able to use the token (or tokens) of the place node. By way of example, it is easy to observe that place node P2 (in other words, adjustment machine M1 in a *free* state), acts as a precondition (in other words, input place node) both for the

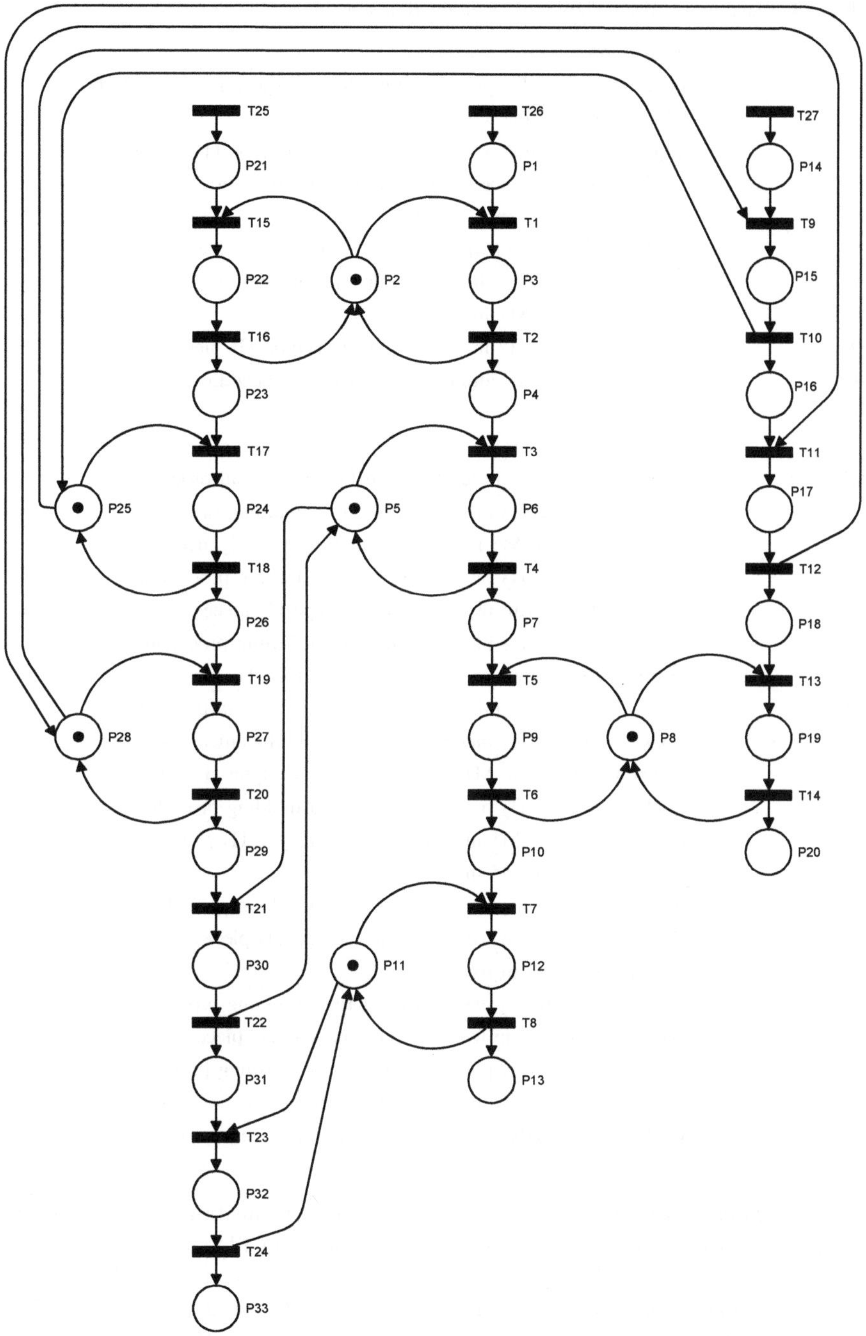

Fig. 3.7 PN process-oriented model for several production sequences

Table 3.4 Place nodes PN system with 3 production sequences

Place	Description
P1	Stock S1 of type Π_1 pieces waiting for an adjustment procedure
P2	Adjustment machine M1 in *free* state
P3	Adjustment machine M1 in *working* state on type Π_1 pieces
P4	Stock S3 of pieces waiting for a pressing procedure
P5	Machine press M3 in *free* state
P6	Machine press M3 in *working* state on type Π_1 pieces
P7	Stock S2 of pieces waiting for a drilling procedure
P8	Drilling machine M2 in *free* state
P9	Drilling machine M2 in *working* state on type Π_1 pieces
P10	Stock S6 of pieces waiting for a polishing procedure
P11	Polisher M6 in *free* state
P12	Polisher M6 in *working* state on type Π_1 pieces
P13	Stock S7 of already processed type Π_1 pieces: end product
P14	Stock S5 of type Π_2 pieces waiting for a molding procedure
P15	Molding machine M5 in *working* state on type Π_2 pieces
P16	Stock S4 of type Π_2 pieces waiting for a milling procedure
P17	Milling cutter M4 in *working* state on type Π_2 pieces
P18	Stock S2 of type Π_2 pieces waiting for a drilling procedure
P19	Drilling machine M2 in *working* state on type Π_2 pieces
P20	Stock S7 of already processed type Π2 pieces: end product
P21	Stock S1 of type Π_3 pieces waiting for an adjustment procedure
P22	Adjustment machine M1 in *working* state on type Π_3 pieces
P23	Stock S5 of type Π_3 pieces waiting for a molding procedure
P24	Molding machine M5 in *working* state on type Π_3 pieces
P25	Molding machine M5 in *free* state
P26	Stock S4 of type Π_3 pieces waiting for a milling procedure
P27	Milling cutter M4 in *working* state on type Π_3 pieces
P28	Milling cutter M4 in *free* state
P29	Stock S3 of type Π_3 pieces waiting for a pressing procedure
P30	Machine press M3 in *working* state on type Π_3 pieces
P31	Stock S6 of type Π_3 pieces waiting for a polishing procedure
P32	Polisher M6 in *working* state on type Π_3 pieces
P33	Stock S7 of already processed type Π_3 pieces: end product

Start adjustment procedure event on a type Π_1 piece (transition T1) and the *Start adjustment procedure* event on a type Π_3 piece (transition T15). In the case of firing transition T1, transition T15 shall be disabled and vice versa. Similarly, also a Decision/Conflict pattern of behavior is presented, but with 2 transitions (T7 and T23) that compete to be able to use the token of place node P11.

Table 3.5 Transition nodes in PN system with 3 production sequences

Transition	Description
T1	Start adjustment procedure on type Π_1 piece
T2	End adjustment procedure on type Π_1 piece
T3	Start pressing procedure on type Π_1 piece
T4	End pressing procedure on type Π_1 piece
T5	Start drilling procedure on type Π_1 piece
T6	End drilling procedure on type Π_1 piece
T7	Start polishing procedure on type Π_1 piece
T8	End polishing procedure on type Π_1 piece
T9	Start molding procedure on type Π_2 piece
T10	End molding procedure on type Π_2 piece
T11	Start milling procedure on type Π_2 piece
T12	End milling procedure on type Π_2 piece
T13	Start drilling procedure on type Π_2 piece
T14	End drilling procedure on type Π_2 piece
T15	Start adjustment procedure on type Π_3 piece
T16	End adjustment procedure on type Π_3 piece
T17	Start molding procedure on type Π_3 piece
T18	End molding procedure on type Π_3 piece
T19	Start milling procedure on type Π_3 piece
T20	End milling procedure on type Π_3 piece
T21	Start pressing procedure on type Π_3 piece
T22	End pressing procedure on type Π_3 piece
T23	Start polishing procedure on type Π_3 piece
T24	End polishing procedure on type Π_3 piece
T25	Arrival of type Π_3 pieces
T26	Arrival of type Π_1 pieces
T27	Arrival of type Π_2 pieces

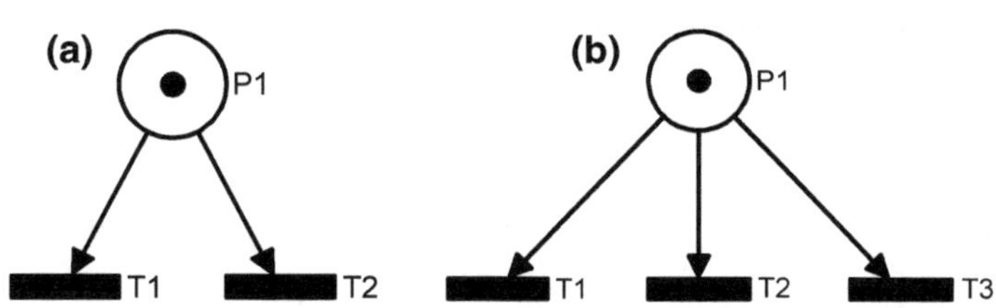

Fig. 3.8 Decision/conflict behavior pattern

Figure 3.8 generically shows the Decision/Conflict pattern of behavior. The left side of the same figure is characterized by a decision-making with 2 alternatives, while on the right side (Fig. 3.8b) there are 3 alternatives.

It is also easy to note that in the "*working* machine" procedures it is necessary to differentiate the type of piece being *worked* on so as to be able to redirect the piece to the next procedure. A similarly thing happens with the stocks which also have to distinguish the type of piece that they store, so different place nodes are required for the same stock.

Another pattern of behavior that is also shown in the PN of Fig. 3.7 is concurrency, which is characterized by two or more changes of state variables in the same instant of time. One example of concurrency is to be found in transition T2 as its firing causes a change of state in machine M1 (changes to be *free*) and at the same time increases by one the number of type Π_1 pieces in stock S3.

Figure 3.9 generically represents the pattern of concurrency behavior. The left side of the same Fig. 3.9 is characterized by a concurrency with 2 changes to the state variables, while on the right side (Fig. 3.9b), there are 3 changes to the state variables.

Finally, we also can see the pattern of synchronization behavior, which is usually utilized when waiting for a resource to continue with the sequence of procedures. One example can be seen in transition T3, that is waiting for a type Π_1 piece to arrive at stock S3 to carry out a pressing procedure. The left side of Fig. 3.10a is characterized by a synchronization of 2 resources, while on the right side (Fig. 3.10b) the synchronization of 3 resources is required.

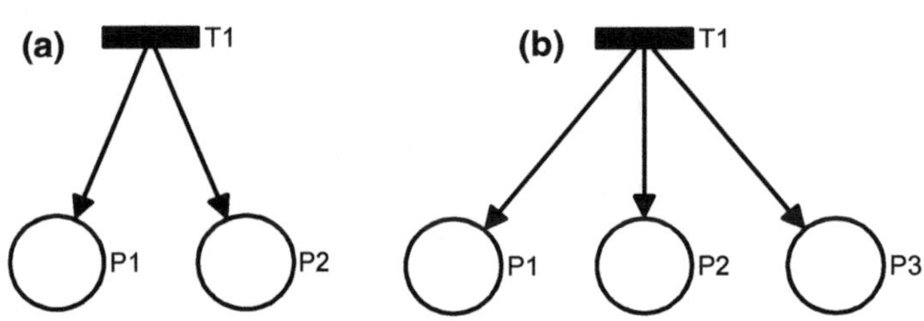

Fig. 3.9 Pattern of concurrency behavior

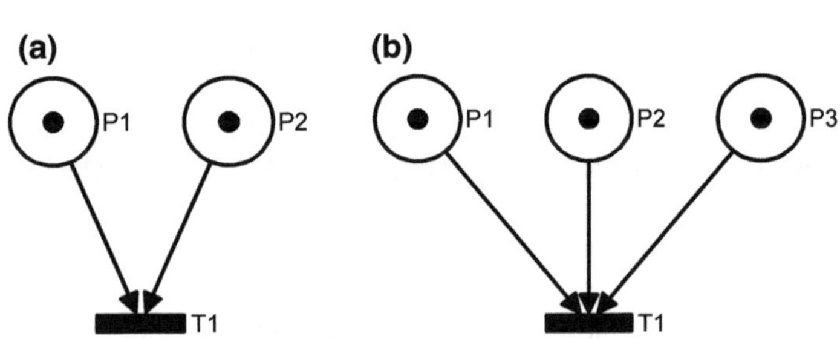

Fig. 3.10 Pattern of synchronization behavior

3.4 Redundant Place Nodes

The purpose is to facilitate the development, maintenance and comprehension of the models, the use of redundant nodes, which can be defined as a duplication of a node that must always maintain the same number of tokens, is admissible in the PN formalism.

Figure 3.11a represents the same PN described in Fig. 3.7, but using 2 redundant nodes for P2, P5, P8, P11, P25 and P28. On the right side of Fig. 3.11, we observe the same PN after having fired transition T1, in which one token has been eliminated from place nodes P1 and P2 and one token has been added to place node P3. It is important to note that in place node P2 connected to the input of transition T15, one token has also been eliminated.

Although colors have been used in Fig. 3.11 to identify the redundant nodes, the only rule there is in this regard is that the place redundant nodes must have the same identifying name.

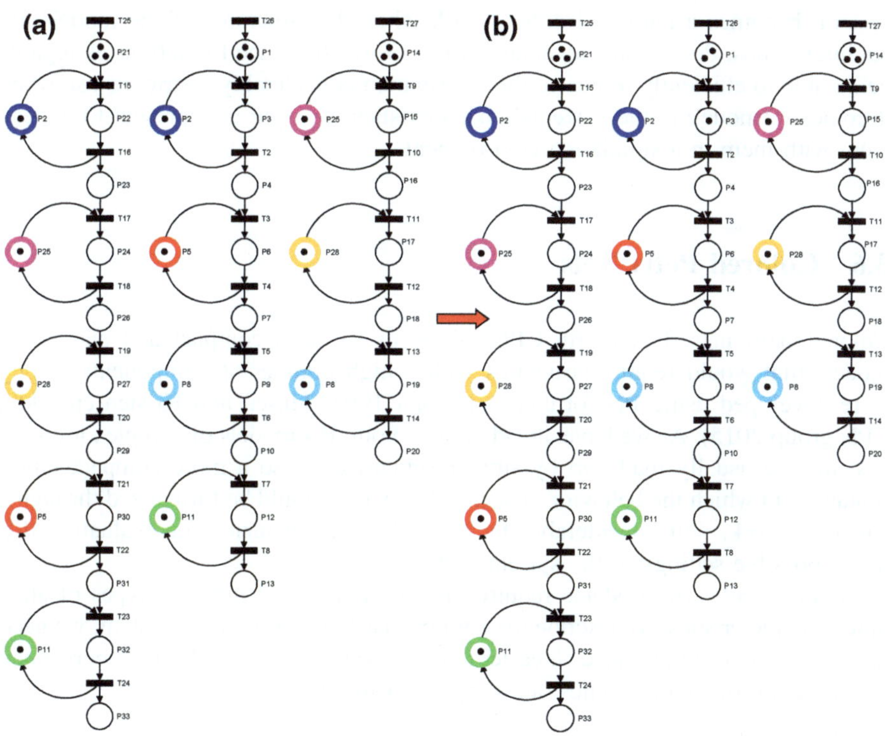

Fig. 3.11 PN model with redundant nodes for several production sequences

3.5 Limitations of Petri Nets

One of the great limitations to the PN formalism for the specification of simulation models for industrial, production, transport and logistics systems, is the inability to describe changes in the information about the entities: not only the resources but also the products that are moved by the production system. The inability to specify information changes in a more compact manner forces the modeler to resort to the specification of similar subnetworks (with the same patterns of behavior and dynamics) where they are only differentiated by the information associated with the place node. In the PN Model represented in Fig. 3.11 it is easy to see that the entire model is the integration of several Petri subnetworks similar to the one represented in Fig. 3.5, where only the information relating to the machine and the stock changes.

The representation of the information in the PN formalism by means of similar subnetworks will result in a considerable rise in the number of place nodes and transition nodes, thus hampering the maintenance and possible exploitation of the developed model (Guasch et al. 2003).

Furthermore, it is worth considering that the task of maintaining the model, which is necessary for the assessment by simulation of different behaviors of the system, becomes a hard and difficult task when the structures of the models are complex. Thus, despite the advantages of PN formalism, the limitations as regards being able to efficiently represent the information considerably restrict its use when it comes to modeling the systems' characteristics we want to represent to experiment with them in a simulation environment.

3.6 Colored Petri Nets

Colored petri nets (henceforth CPN) allow us to build compact and parametric models that would require structures with a high number of components if they were developed using PN formalism (Jensen 1997; Jensen and Kristensen 2009; CPN group 2015). As we have already pointed out, PN models of complex logistics systems are usually made up by similar subnetworks, so a more compact representation, in which the behavior of such subnetworks could be formalized through a single network, will considerably facilitate their maintenance for evaluating different possible settings of the system under study.

As most industrial systems require the specification of different types of attributes (characteristics) to describe the entities that flow through the system, we need to use a formalism that makes it easier to describe the flow of tasks to be performed as a function of said attributes, allowing us to stipulate:

- The priority of the entities in a queue.
- How an event can affect the values of the attributes of the entity being processed (state of the entity).
- Activation of events as a function of the attributes of each entity.

- Length of time of an activity as a function of the attributes of the entities involved.
- The flow of entities and what happens to each entity (changes of state of the entity) as it flows through a sequence of physical subsystems.
- The changes of state of the system and the sequence of events that would have to appear for an entity to finish a particular type of processing.

CPNs support a certain level of abstraction in the modeling stage by using colors to represent the attributes of the entities and which are supported by the majority of commercial simulation software packages.

3.6.1 Elements Involved in the Modeling of Colored Petri Nets

The main difference made by CPNs in respect of ordinary PNs is the ability to associate a type of data (set of values) known as *color* token with every entity (token). The value of a datum can be of an arbitrarily complex type; for example, a log where the first field is a real number, the second is a *string* text, while the third could be a list of whole numbers.

For a given place node, all the tokens must have colors that belong to the specified type. This type is called the *color set* of the place node. The use of color sets in CPN is completely analogous to the use of types in programming languages, which gives them the necessary power to be able to formalize the model of any system, no matter how complex.

Color sets determine the possible values of tokens, in a similar way to how types determine the possible values of variables and expressions. For historical reasons we talk about colored tokens that are distinguished from one another, in contrast to the flat tokens of ordinary PNs. However, we can also talk about values and types instead colors and color sets.

Apart from the basic elements of a PN described in Sect. 3.2 (place nodes, transition nodes and arcs), the concepts of color and color set introduce into a CPN the use of the following representation elements (Jensen 1997):

- **Color sets:** each place node can only have tokens with the same type of data, which is known as a color set of the place node. This constraint is totally compatible with the modeling objectives, always provided the place nodes represent either conditions or queues. This is graphically represented with the name of the color set to one of the place node.
- **Initialization expressions:** indicate the initial number of tokens and their color in each one of the place nodes. Graphically speaking, the initialization expressions express the number of tokens in a place node with a number in a circle beside the node. The colors of the tokens is specified by means of an underlined expression beside the place node with the following information:

$$\underline{n'(c_1, c_2, c_3, \ldots, c_r, \ldots, c_{nr})}$$

where:

n: represents the number of tokens with the color values described inside the brackets.
c_r: represents the value of a color attribute.
nr: represents the number of color attributes of the tokens.

When the values of the colors of the tokens are not identical for all the elements of the same place node, the operator "+" is used to specify the values of the colors of each token:

$$n1'(c_{11},\ c_{12},\ c_{13}, \ldots,\ c_{1r}, \ldots,\ c_{1nr}) + n2'(c_{21},\ c_{22},\ c_{23}, \ldots,\ c_{2r}, \ldots,\ c_{2nr})$$

- **Initial state:** the initial state (marking M0) shall be determined by assessing the initialization expressions associated with each place node, which shall determine the number of tokens in each place node, as well as the values of the colors of the tokens.
- **Arc expressions:** the colors of the tokens can be inspected in the transitions, which will make it possible to enable them not only according to the number of tokens in the place nodes connected to the input of the transition, but also according to the color values of the tokens available in said place node and, at the same time, will also make it possible to model the effects of each transition, defining new colors for the output tokens.
  Arc expressions consist of the formalization of constraints between the colors of the various tokens of the place nodes connected to the input of the transition, for which formalization variables that have been assigned specific token color values can be used to force a selection of those tokens whose colors coincide with the values of said variables.
- **Guards:** guards have a similar function to the arc expressions, but are only logical expressions (*Booleans*) that impose certain values on the colors of the tokens that can be chosen to enable a transition. They are graphically formalized between square brackets "[ ]" placed beside the transition.
- **Marking:** represents the minimum information required to be able to predict what possible events could be produced. It is described through the specification of the number of tokens in each place node, as well as the values of the colors of every one of the tokens.

3.6.2 Formal Definition of Colored Petri Nets

In formal terms, a CPN is defined as a tuple (Jensen 1997; Jensen and Kristensen 2009; Guasch et al. 2003; Narciso et al. 2010):

$$CPN = (\Sigma, P, T, A, N, C, G, E, I)$$

where:

Σ = {C1, C2, …, Cnc}:	Finite and non-empty color sets (*nc* is the number of color sets specified for the CPN).
P = {P1, P2, P3, …, Pnp}:	Finite set of place nodes (*np* is the number of place nodes in the CPN).
T = {T1, T2, T3, …, Tnt}:	Finite set of transition nodes (*nt* is the number of transition nodes in the CPN).
A = {a1, a2, a3, …, ana}:	Finite set of arcs that connect nodes *P* with nodes *T* and vice versa (*na* is the number of arcs in the CPN).

N: **Node** function, N: A → (P × T) ∪ (T × P), that makes it possible to associate its terminal nodes with each arc in the form of an ordered pair, so that:

$$\forall \mathrm{Ai} \in \mathrm{A}\, \exists!\, \mathrm{Pj} \in \mathrm{P} \wedge \exists!\, \mathrm{Tk} \in \mathrm{T} : [\mathrm{N(Ai)} = (\mathrm{Pj}, \mathrm{Tk}) \vee \mathrm{N(Ai)} = (\mathrm{Tk}, \mathrm{Pj})]$$

where the first element of the ordered pair corresponds to the origin node and the second element to the destination node. The two nodes have to be of different types, accordingly, if one node is a transition node the other must be a place node and vice versa.

C: **Color** function, C: P → Σ, that makes it possible to specify, for each place node, the type of entities (tokens) that can stored, so that:

$$\forall \mathrm{P_j} \in \mathrm{P}\, \exists!\, \mathrm{Cq} \in \Sigma : [\mathrm{C(Pj)} = \mathrm{Cq}]$$

G: **Guard** function, that allows us to associate each transition node with a logical expression, G: T → *Boolean*, so that:

$$\forall \mathrm{Tk} \in \mathrm{T} : [\mathrm{type(G(Tk))} = Boolean \wedge \mathrm{type(variables(G(Tk))} \subseteq \Sigma]$$

E: **Arc expression** function, E: A → C(Pj) that makes it possible to specify the type of entity (token) of the input place node to a transition that must be chosen from among the tokens stored in said node to enable the transition, so that:

$$\forall \mathrm{Ai} \in \mathrm{A} : [\mathrm{type(E(Ai))} = \mathrm{C(Pj)} \wedge \mathrm{type(variables(E(Ai)))} \subseteq \Sigma]$$

where Pj represents the input or output place node from arc Ai.
When expression E is found associated with an output arc of the transition, the expression is used to assess the new color values of the attributes of the output entities (tokens).

I: **Initialization** function, P → C(Pj), that makes it possible to specify the color values of the attributes of the entities (tokens) initially stored in a place node, so that:

$$\forall P_j \in P : [\text{type}(I(Pj)) = C(Pj)]$$

Each one of these components of a CPN makes it possible to specify and/or represent any component of a simulation model for a DES, such as:

- The Σ set enables us to specify, for each type of entity to be modeled, the attributes that must be defined in the code of the simulation model.
- Each place node can represent, for example, one or more production units (machines) with one or more queues.
- The transitions in the model correspond to events that are usually encoded, such as the start or end of a certain activity,[4] or else the end of an external event as in the case of an arrivals process.
- Guard functions can be used to disinhibit the event associated with a transition as a function of the values of the attributes of an entity to be processed.

So CPNs provide us with the necessary knowledge representation tools to be able to formalize not only the attributes or characteristics of the entities that flow in the system, but also the properties that these entities need to have so that a certain event can happen, and said tools have demonstrated that they are the right ones for modeling logistical systems thanks to their various advantages, such as the ability to contain not only the static structure but also the dynamics of the system, the architecture of the system, and its graphic characteristics (Silva and Valette 1989; Zimmermann et al. 1996; Jensen 1997; Jensen and Kristensen 2009; Piera et al. 2009; Narciso 2010; Narciso and Piera 2015).

3.6.3 *Behavior or Dynamics of Colored Petri Nets*

As with the PN, the arc expressions indicate the necessary conditions for a transition to be activated. However, in a CPN it is not just enough for a place node to contain the number of tokens specified in the arc that goes from the place node to the transition, but it must also contain the color of the tokens that will enable said transition, while the condition expressed by the guard associated with the transition

[4]As with the PNs, depending on the abstraction made of the system to be modeled, an activity can be represented in a CPN as a place node or a transition node.

should also, when applicable, be satisfied. For example, let us consider the CPN that is formalized in Fig. 3.12, which represents a production machine with an incoming stock (place node P1), and 2 ongoing stocks where the type Π_1 pieces (place node P3) are stored and the type Π_2 pieces (place node P2). Place nodes P4 and P5 are used to indicate the *working* and *free* state of the machine respectively. Transition T1 indicates the "Start production operation" event, while event T2 represents the "End production operation on piece Π_1", and event T2 represents "End production operation on piece Π_2". Initially there are 3 type Π_1 pieces and three type Π_2 pieces in the stock coming into the machine that are represented by the initialization expression 3′(1) + 3′(2) located at the side of node P1, and the machine is in a *free* state that is presented by the initialization expression 1′(1) located at the side of place node P5. The rest of the place nodes do not have any token.

The initial marking is represented mathematically by the following vector:

$$M0 = [3'(1) + 3'(2), \ldots, 1'(1)]$$

where the tokens of each place node are delimited by the symbol ",". Thus, we can observe that tokens 3′(1) + 3′(2) correspond to place node P1 and token 1′(1) corresponds to place node P5, while there is no token in place nodes P2, P3 and P4.

For transition T1 in Fig. 3.12 to be enabled it is necessary for the following conditions to be fulfilled:

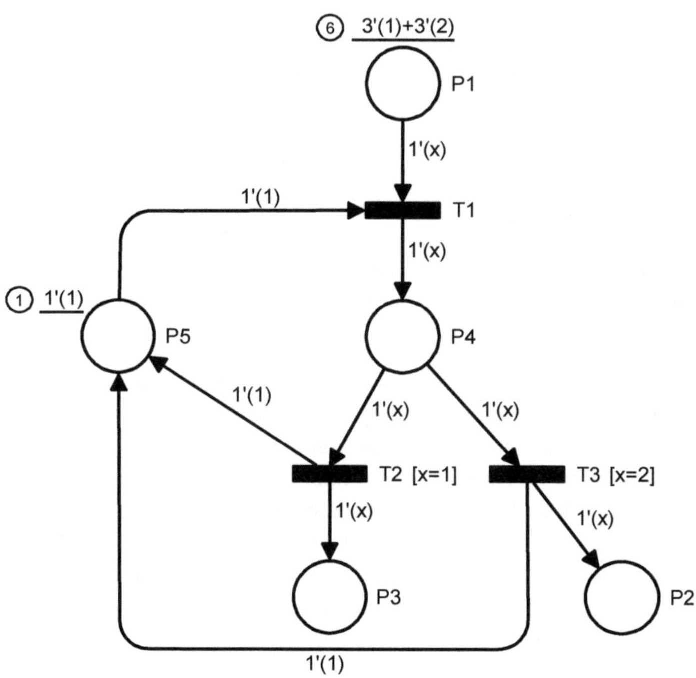

Fig. 3.12 CPN model of a production machine with 2 types of pieces

- There must be at least one token with any color in place node P1. Arc expression $1'(x)$ indicates one token whose color will be instantiated in variable x. Thus, if a color 1 token (in other words, 1′(1)) is chosen, variable x will be instantiated at value 1. Whereas, if a color 2 token (in other words, 1′(2)) is chosen, variable x will be instantiated at value 2.
- In place node P5 there must be at least one token with the color 1 (arc expression 1′(1)).

As an effect of firing transition T1 one token is eliminated from place node P1 and one token is eliminated from place node P5, and a new token shall be added to place node P4 with the value of variable x. Thus, in the case of firing transition T1 with one token of color 1 (in other words, one type Π_1 piece) from place node P1, the new state that would be achieved is represented by vector M1. If a type Π_2 piece (in other words, token of color 2 of place node P1) is chosen, the new state that would be achieved is represented by vector M2.

$$\begin{aligned} M1 &= [2'(1) + 3'(2), \ldots, 1'(1),] \\ M2 &= [3'(1) + 2'(2), \ldots, 1'(2),] \end{aligned}$$

Given state M1, transition T2 is enabled because the guard associated with said transition demands that variable x takes a value of 1 (note that there is only one token with value 1 in place node P4). Similarly, given state M2, transition T3 is found enabled as the guard associated with said transition demands that variable x take a value of 2. Vector M3 corresponds to the state that is reached if transition T2 is fired from state M1, while vector M4 corresponds to the state that is reached if transition T3 is fired from state M2.

$$\begin{aligned} M3 &= [2'(1) + 3'(2), , 1'(1), \ldots, 1'(1)] \\ M4 &= [3'(1) + 2'(2), 1'(2), \ldots, 1'(1)] \end{aligned}$$

In this sense, the CPN formalism provides a formal method that permits us to represent and fully assess its behavior. This method is known as the CPN *scope tree* or occurrence graph (Jensen 1997; Narciso 2010).

Example 3.3 CPN Model of different sequences of production operations on different types of pieces while maintaining a flow of products

Let us consider that, in this example, we have to process 3 different types of pieces, each one with a sequence of production operations set forth in the recipe described in Table 3.3. Table 3.6 gives the information about the place nodes, Table 3.7 defines the colors used and Table 3.8 represents the information about the transition nodes, while Table 3.9 gives the information about the arc expressions.

Figure 3.13 graphically represents the CPN of the various sequences of production operations on the 3 types of pieces.

Initially all the place nodes are empty except for the place nodes that represent the machines in *free* state, whose tokens are:

Table 3.6 Process-oriented CPN place nodes for a system with 3 production sequences

Place	Color	Description
P1	C1	Stock S1 of type Π_1, Π_2 and Π_3 pieces, waiting the first production operation
P2	C2	Adjustment machine M1 in *free* state
P3	C2	Adjustment machine M1 in *working* state on type Π_1 pieces
P4	C1	Stock S3 of pieces waiting for a pressing procedure
P5	C2	Machine press M3 in *free* state
P6	C2	Machine press M3 in *working* state on type Π_1 pieces
P7	C1	Stock S2 of pieces waiting for a drilling procedure
P8	C2	Drilling machine M2 in *free* state
P9	C2	Drilling machine M2 in *working* state on type Π_1 pieces
P10	C1	Stock S6 of pieces waiting for a polishing procedure
P11	C2	Polisher M6 in *free* state
P12	C2	Polisher M6 in *working* state on type Π_1 pieces
P13	C1	Stock S7 of finished type Π_1, Π_2, and Π_3 pieces
P14	C2	Molding machine M5 in *free* state
P15	C2	Molding machine M5 in *working* state on type Π_2 pieces
P16	C1	Stock S4 of type Π_2 pieces waiting for a milling procedure
P17	C2	Milling cutter M4 in *working* state on type Π_2 pieces
P18	C1	Stock S2 of type Π_2 pieces waiting for a drilling procedure
P19	C2	Drilling machine M2 in *working* state on type Π_2 pieces
P20	C2	Adjustment machine M1 in *working* state on type Π_3 pieces
P21	C1	Stock S5 of type Π_3 pieces waiting for a molding procedure
P22	C2	Modeling machine M5 in *working* state on type Π_3 pieces
P23	C1	Stock S4 of type Π_3 pieces waiting for a milling procedure
P24	C2	Milling cutter M4 in *working* state on type Π_3 pieces
P25	C2	Milling cutter M4 in *free* state
P26	C1	Stock S3 of type Π_3 pieces waiting for a pressing procedure
P27	C2	Machine press M3 in *working* state on type Π_3 pieces
P28	C1	Stock S6 of type Π_3 pieces waiting for a polishing procedure
P29	C2	Polisher M6 in *working* state on type Π_3 pieces

Table 3.7 Definition of colors C1 and C2

Color	Definition	Description
C1	Integer 1…3	Identification of type of piece
C2	Integer 1…6	Identification of machine or stock

$$P2 = 1'(1);\quad P5 = 1'(3);\quad P8 = 1'(2);\quad P11 = 1'(6);\quad P14 = 1'(5);\quad P25 = 1'(4)$$

Unlike the production system described in Example 3.2, in this system the arrivals process of type Π_1, Π_2, Π_3 pieces has been modeled by means of events T1, T2 and T3 respectively. As can be seen in the CPN of Fig. 3.13, these events

Table 3.8 Process-oriented CPN transition nodes for a system with 3 production sequences

Transition	Description
T1	Arrival of type Π_1 piece
T2	Arrival of type Π_2 piece
T3	Arrival of type Π_3 piece
T4	Start adjustment procedure on type Π_1 piece
T5	End adjustment procedure on type Π_1 piece
T6	Start pressing procedure on type Π_1 piece
T7	End pressing procedure on type Π_1 piece
T8	Start drilling procedure on type Π_1 piece
T9	End drilling procedure on type Π_1 piece
T10	Start polishing procedure on type Π_1 piece
T11	End polishing procedure on type Π_1 piece
T12	Start molding procedure on type Π_2 piece
T13	End molding procedure on type Π_2 piece
T14	Start milling procedure on type Π_2 piece
T15	End milling procedure on type Π_2 piece
T16	Start drilling procedure on type Π_2 piece
T17	End drilling procedure on type Π_2 piece
T18	Start adjustment procedure on type Π_3 piece
T19	End adjustment procedure on type Π_3 piece
T20	Start molding procedure on type Π_3 piece
T21	End molding procedure on type Π_3 piece
T22	Start milling procedure de of type Π_3 piece
T23	End milling procedure on type Π_3 piece
T24	Start pressing procedure on type Π_3 piece
T25	End pressing procedure on type Π_3 piece
T26	Start polishing procedure on type Π_3 piece
T27	End polishing procedure on type Π_3 piece

Table 3.9 Process-oriented CPN arc expressions for a system with 3 production sequences

Arc	Expression
a1, a4, a5, a6, a7, a8, a9, a10, a12, a13, a15, a16, a18, a19, a21, a22, a24, a25, a27, a48, a50	1′(1)
a2, a17, a20, a28, a30, a31, a33, a34, a36, a37, a38, a40, a41, a42, a43, a44, a45	1′(2)
a3, a11, a14, a46, a47, a49, a51, a52, a54, a56, a57, a59, a60, a62, a63, a64, a65, a66, a67, a68, a69, a71, a72, a74, a75	1′(3)
a35, a39, a58, a61	1′(4)
a29, a32, a53, a55	1′(5)
a23, a26, a70, a73	1′(6)

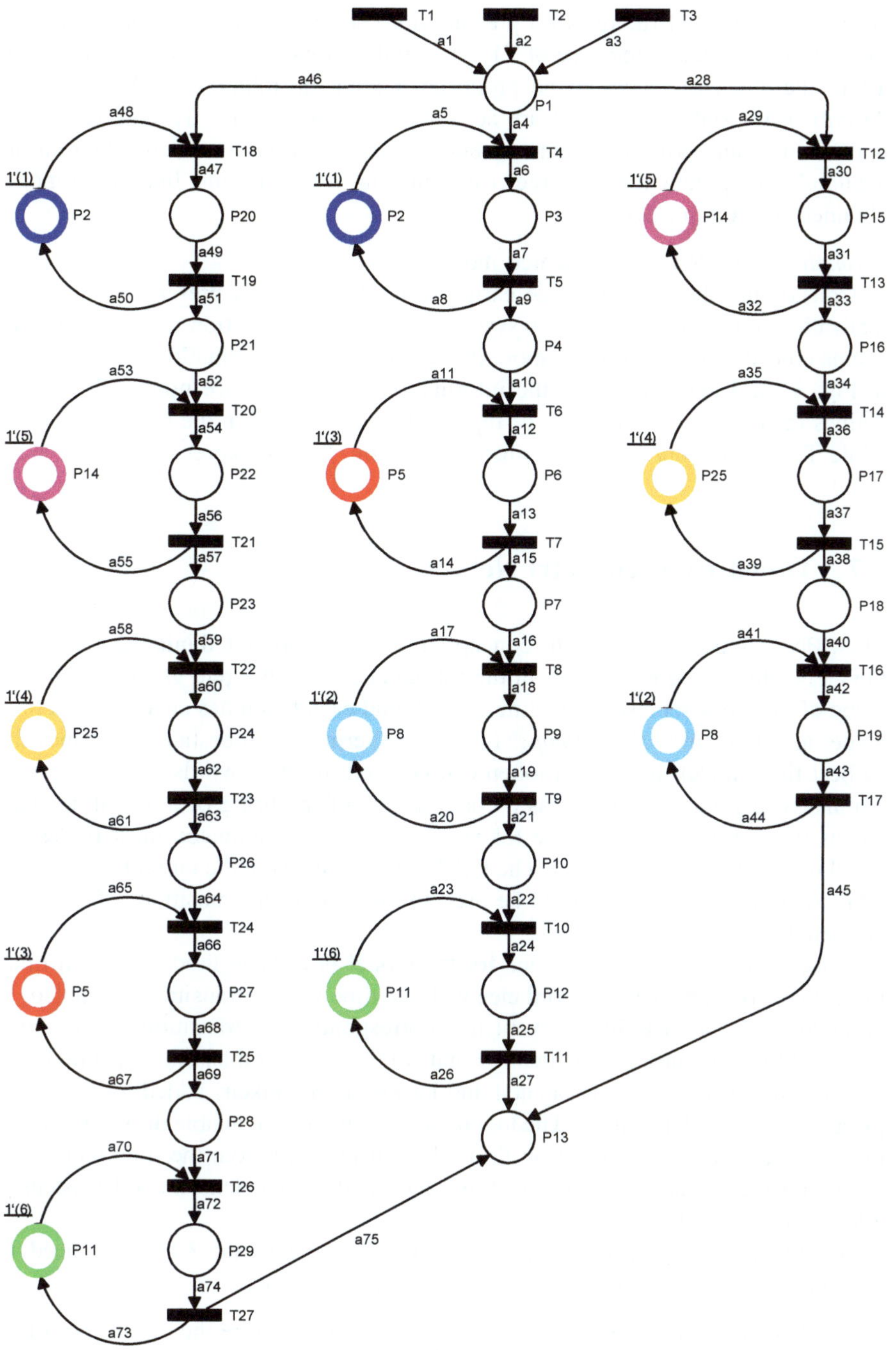

Fig. 3.13 Process-oriented CPN model

are always activated (there is no precondition) so, every time they are fired, one token 1′(1) is added in the case of T1, 1′(2) in the case of T2, or 1′(3) in the case of T3 to the stock of pieces to be produced. Given the initial state that has been described, the only events that are activated are precisely T1, T2 and T3 which, once fired, shall permit different sequences of events to be fired depending on the availability of production resources (machines in *free* state), just like in the model obtained in Example 3.2.

Example 3.4 CPN Layout-oriented model

Considering the system in the earlier example, Fig. 3.14 presents another formalization of the same production system at a level of abstraction where the flow of events considers a layout orientation, instead of a process orientation as represented in Fig. 3.13. Table 3.10 gives the information about the place nodes, Table 3.11 defines the colors used, Table 3.12 represents the information the transition nodes, while Table 3.13 represents the information about the arc expressions.

3.7 Timed Colored Petri Nets

The CPNs can be extended if the concept of time is incorporated into the models. In the CPNs this concept is based on the introduction of a discrete *global clock*. The value of the clock represents the time model, and each token has an associated time value, also known as *time stamps*. In intuitive terms, the timestamp describes the earliest time model in which a token can be used, in other words, when it can be eliminated from a place node, as a consequence of the firing of a transition. This means that the timestamp of the tokens that are to be eliminated must be less or equal to the current time model. The CPN model remains in the current time model while there are tokens that could be used to fire any of the transitions of the CPN (Jensen 1997).

Once all the possible transitions for the model have been fired in current time, the global clock updates in accordance with the time when a transition is enabled. In order to model an activity or event that corresponds to a transition that requires *r* units of time, a timestamp, which is *r* units of time greater than the model in which the transition was fired is associated and located in the tokens added to the output place nodes of said transition. The tokens will not then be available during *r* units of time and cannot be eliminated by firing the transitions, before the time model has been increased by at least *r* units of time (Jensen 1997; Kristensen and Christensen 2004; Narciso et al. 2012).

So, in order to incorporate the concept of time into the CPN, we need to introduce the following modeling elements into this model:

- A *global clock* that represents the time immediately before the occurrence of an event or the firing of a transition.
- A *time value* (timestamp) associated with each token in a marking, that describes the smallest time value where a token can be used to enable a

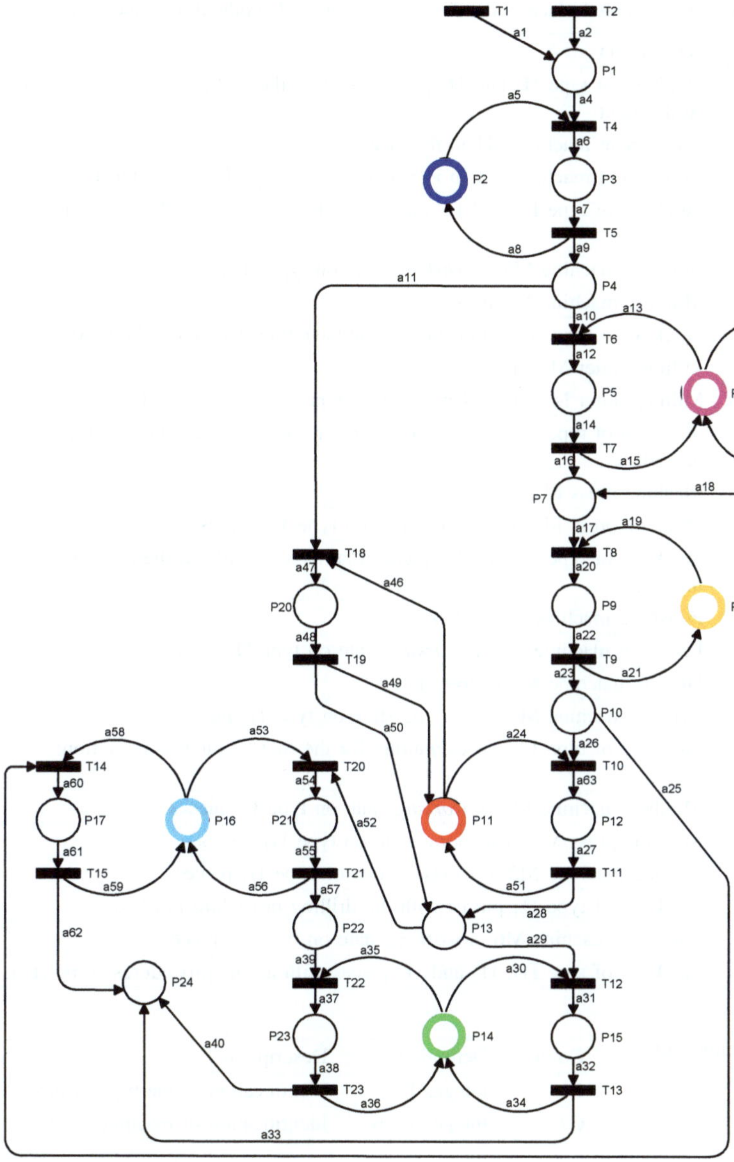

Fig. 3.14 Layout-oriented model

transition, in other words, it indicates when the token is "ready" to be used by a transition.

- A *delay time* (*r*) associated with a transition, that indicates, for example, that an activity modeled through a transition consumes time. In this chapter the term *transition time* is used to represent this concept.

Table 3.10 CPN layout-oriented place nodes for a system with 3 production sequences

Place	Color	Description
P1	C1	Stock S1 of type Π_1 and Π_3 pieces, waiting the first production operation: Adjustment
P2	C2	Adjustment machine M1 in *free* state
P3	C1	Adjustment machine M1 in *working* state on type Π_1 or Π_3 pieces
P4	C1	Stock S3 of type Π_1 and Π_3 pieces that have completed the adjustment procedure
P5	C1	Molding machine M5 in *working* state on type Π_3 pieces
P6	C2	Molding machine M5 in *free* state
P7	C1	Stock S2 of type Π_3 and Π_2 pieces that have finished the molding procedure
P8	C2	Milling cutter M4 in *free* state
P9	C1	Milling cutter M4 in *working* state on type on type Π_2 or Π_3 pieces
P10	C1	Stock S6 of type Π_2 and Π_3 pieces that have completed the milling procedure
P11	C2	Machine press M3 in *free* state
P12	C1	Machine press M3 in *working* state on type Π_3 pieces
P13	C1	Stock S7 of type Π_1 and Π_3 pieces that have completed the pressing procedure
P14	C2	Polishing machine M6 in *free* state
P15	C1	Polishing machine M6 in *working* state on type Π_3 pieces
P16	C2	Drilling machine M2 in *free* state
P17	C1	Drilling machine M2 in *working* state on type Π_2 pieces
P18	C1	Stock S3 of type Π_2 pieces waiting for the first production operation: molding
P19	C1	Molding machine M5 in *working* state on type Π_2 pieces
P20	C1	Machine press M3 in *working* state on type Π_1 pieces
P21	C1	Drilling machine M2 in *working* state on type Π_1 pieces
P22	C1	Stock S5 of type Π_1 pieces with the drilling procedure finished
P23	C1	Polishing machine M6 in *working* state on type Π_1 pieces
P24	C1	Stock S4 of type Π_1, Π_2 and Π_3 pieces with all the procedures completed

Table 3.11 Definition of colors C1 and C2

Color	Definition	Description
C1	Integer 1…3	Identification of the type of piece
C2	Integer 1…6	Identification of machine or stock

- An *arrival time* of a marking that represents the time it takes the firing of a transition to change the state of the system to a new state.

In this case, a transition will not only be enabled when the tokens satisfy the corresponding arc input expressions of the transition, but these tokens must also be "ready". This means that all the times (timestamps) associated with the tokens that enable the transition, should be shorter than or equal to the current value of the

Table 3.12 CPN layout-oriented transition nodes for a system with 3 production sequences

Transition	Description
T1	Arrival of type Π_1 piece
T2	Arrival of type Π_3 piece
T3	Arrival of type Π_2 piece
T4	Start procedure to adjust type Π_1 or Π_3 piece
T5	End adjustment procedure on type Π_1 or Π_3 piece
T6	Start molding procedure on type Π_3 piece
T7	End molding procedure on type Π_3 piece
T8	Start milling procedure on type Π_2 or Π_3 piece
T9	End milling procedure on type Π_2 or Π_3 piece
T10	Start pressing procedure on type Π_3 piece
T11	End pressing procedure on type Π_3 piece
T12	Start polishing procedure on type Π_3 piece
T13	End polishing procedure on type Π_3 piece
T14	Start drilling procedure on type Π_2 piece
T15	End drilling procedure on type Π_2 piece
T16	Start molding procedure on type Π_2 piece
T17	End molding procedure on type Π_2 piece
T18	Start pressing procedure on type Π_1 piece
T19	End pressing procedure on type Π_1 piece
T20	Start drilling procedure on type Π_1 piece
T21	End drilling procedure on type Π_1 piece
T22	Start polishing procedure on type Π_1 piece
T23	End polishing procedure on type Π_1 piece

Table 3.13 CPN layout-oriented arc expressions for a system with 3 production sequences

Arc	Expression
a4, a6, a7, a9, a17, a20, a22, a23	1′(x)
a1, a5, a8, a11, a37, a38, a39, a40, a47, a48, a50, a52, a54, a55, a57	1′(1)
a3, a18, a25, a42, a43, a44, a45, a53, a56, a58, a59, a60, a61, a62	1′(2)
a2, a10, a12, a14, a16, a24, a51, a26, a63, a27, a28, a29, a31, a32, a33, a46, a49	1′(3)
a19, a21	1′(4)
a13, a15, a41, a45	1′(5)
a30, a34, a35, a36	1′(6)

clock. Otherwise, the global clock must advance to the lowest timestamp value of the tokens for which the transition would be enabled.

To illustrate how the timed CPNs behave, let us look again at the CPN in Fig. 3.12, and assume:

- A global clock initialized at 0.
- A delay time of 10 units associated with transition T2 that indicates the time required to complete a production operation on a type Π_1 piece.
- A delay time of 4 units associated with transition T3 that indicates the time required to complete a production operation on a type Π_2 piece.
- Transition T1 does not consume time.

The symbol @ is used to indicate the information about time. So, if we assume that for the initial marking all the tokens have associated times that are equal to zero, M0 can be represented, with its corresponding timing information:

$$M0 = [3'(1)\ @0 + 3'(2)\ @0, \ldots, 1'(1)\ @0]$$

The three tokens of value 1 in place node P1 have a token time 0. The same information is read in the rest of the tokens. If we consider that the tokens of value 1 in P1 arrived in the instants of time 2, 4 and 6, the marking would be represented by the following vector:

$$M0 = [1'(1)\ @2 + 1'(1)\ @4 + 1'(1)\ @6 + 3'(2)\ @0, \ldots, 1'(1)\ @0]$$

Figure 3.1 represents the timed RdPC.

Below the evolution of the state of the system is represented, considering the firing sequence of transitions T1:1′(2) − T3 − T1:1′(1) − T2 − T1:1′(1), after initial state M0 represented in Fig. 3.15 and clock time 0.00:

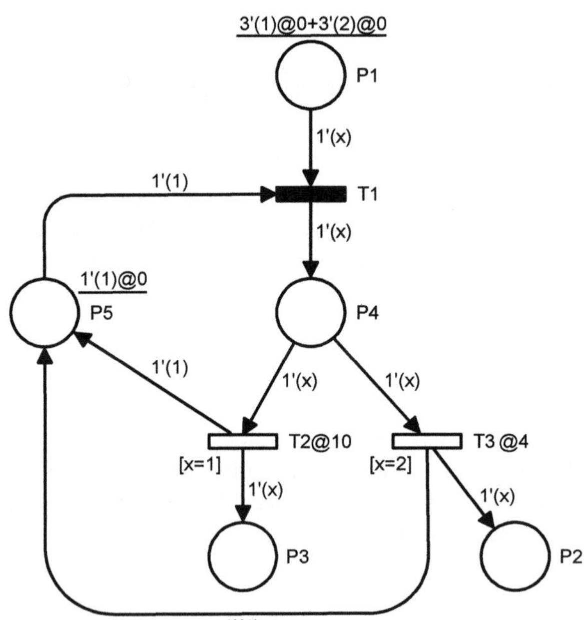

Fig. 3.15 Timed CPN model of a production machine with 2 types of pieces

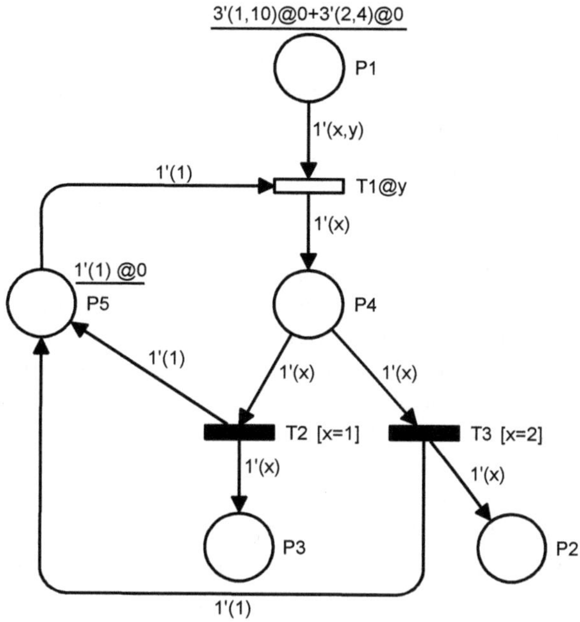

Fig. 3.16 Timed CPN model of a production machine with 2 types of pieces, using color to specify the time

$$
\begin{aligned}
\text{M0} &= [3'(1)\,@0 + 3'(2)\,@0, \ldots, 1'(1)\,@0]\ \text{Clock} : 0.00\\
\text{M1} &= [3'(1)\,@0 + 2'(2)\,@0, \ldots, 1'(2)\,@0,]\ \text{Clock} : 0.00\\
\text{M2} &= [3'(1)\,@0 + 2'(2)\,@0, 1'(2)\,@4, \ldots, 1'(1)\,@4]\ \text{Clock} : 0.00\\
\text{M3} &= [2'(1)\,@0 + 2'(2)\,@0, 1'(2)\,@4, \ldots, 1'(1)\,@4,]\ \text{Clock} : 4.00\\
\text{M4} &= [2'(1)\,@0 + 2'(2)\,@0, 1'(2)\,@4, 1'(1)\,@14, \ldots, 1'(1)\,@14]\ \text{Clock} : 4.00\\
\text{M5} &= [1'(1)\,@0 + 2'(2)\,@0, 1'(2)\,@4, 1'(1)\,@14, 1'(1)\,@14,]\ \text{Clock} : 14.00
\end{aligned}
$$

With the incorporation of timing into the models, it is possible to use the CPN to assess not only the behavior but also the performance of a system (Fig. 3.16).

Figure 3.1 represents the same system, but the time in transition T1 has been associated, using the second color attribute of the tokens in place node P1 (variable *y*), which represents the time required by each type of piece to complete the corresponding production operation. The dynamics of the system obtained are exactly the same but with a different representation of the information, in that the time of the tokens increases at the start of the production operation and not at the end but it keep the same evolution of the simulation clock, as can be seen in the evolution of the states

$$
\begin{aligned}
\text{M0} &= [3'(1,10)@0 + 3'(2,4)@0, \ldots, 1'(1)@0] \text{ Clock : } 0.00 \\
\text{M1} &= [3'(1,10)@0 + 2'(2,4)@0, \ldots, 1'(2)@4,] \text{ Clock : } 0.00 \\
\text{M2} &= [3'(1,10)@0 + 2'(2,4)@0, 1'(2)@4, \ldots, 1'(1)@4] \text{ Clock : } 4.00 \\
\text{M3} &= [2'(1,10)@0 + 2'(2,4)@0, 1'(2)@4, \ldots, 1'(1)@14,] \text{ Clock : } 4.00 \\
\text{M4} &= [2'(1,10)@0 + 2'(2,4)@0, 1'(2)@4, 1'(1)@14, \ldots, 1'(1)@14] \text{ Clock : } 14.00 \\
\text{M5} &= [1'(1,10)@0 + 2'(2,4)@0, 1'(2)@4, 1'(1)@14, 1'(1)@24,] \text{ Clock : } 14.00
\end{aligned}
$$

CPN formalism also enables us to specify the consumption of time in the place nodes so that they can represent activities. Figure 3.1 represents the same system associating the timed activity with place node P4, which uses the second color attribute (variable *y*) that indicates the time that each piece requires to complete the production operation (Fig. 3.17).

The evolution of the states is shown below:

$$
\begin{aligned}
\text{M0} &= [3'(1,10)\,@0 + 3'(2,4)\,@0, \ldots, 1'(1)\,@0] \text{ Clock : } 0.00 \\
\text{M1} &= [3'(1,10)\,@0 + 2'(2,4)\,@0, \ldots, 1'(2,4)\,@0,\,] \text{ Clock : } 0.00 \\
\text{M2} &= [3'(1,10)\,@0 + 2'(2,4)\,@0, 1'(2)\,@4, \ldots, 1'(1)\,@4] \text{ Clock : } 0.00 \\
\text{M3} &= [2'(1,10)\,@0 + 2'(2,4)\,@0, 1'(2)\,@4, \ldots, 1'(1,10)\,@4,\,] \text{ Clock : } 4.00 \\
\text{M4} &= [2'(1,10)\,@0 + 2'(2,4)\,@0, 1'(2)\,@4, 1'(1)\,@14,, 1'(1,10)\,@14] \text{ Clock : } 4.00 \\
\text{M5} &= [1'(1,10)\,@0 + 2'(2,4)\,@0, 1'(2)\,@4, 1'\,(1)\,@14, 1'(1,10)\,@24,\,] \text{ Clock : } 14.00
\end{aligned}
$$

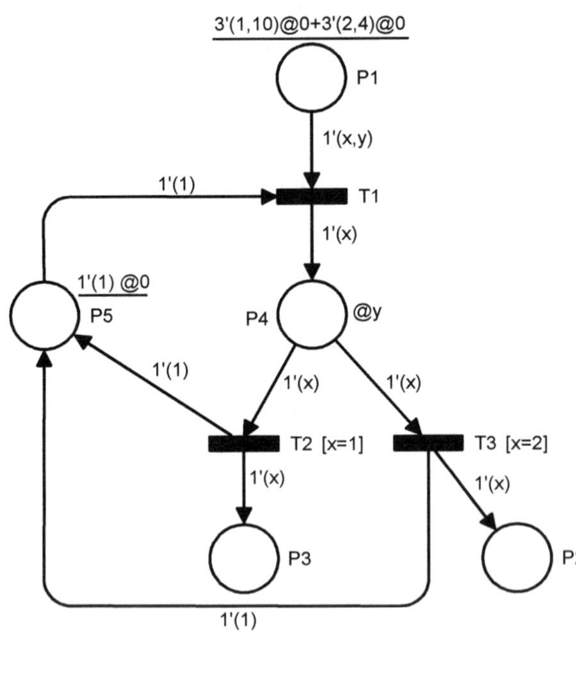

Fig. 3.17 Timed CPN model for a production machine with 2 types of pieces with the time associated with the place node

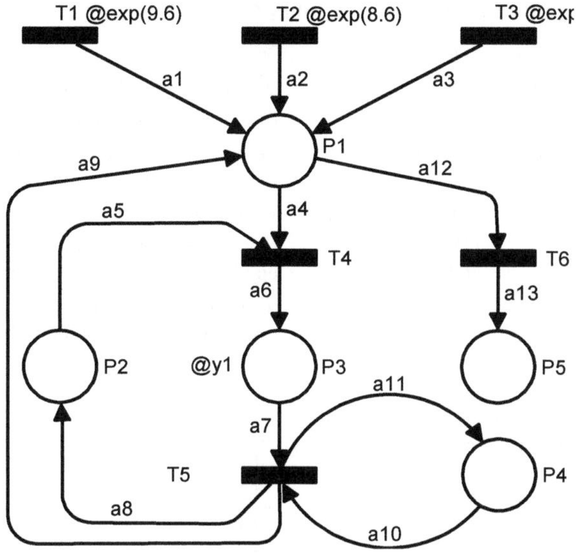

Fig. 3.18 Timed CPN model for a production system with 3 types of pieces

Example 3.5 Timed CPN model of different sequences of production operations on different types of pieces

Figure 3.18 represents the system described in Example 3.3, but with timed representation, where the pieces arrive at the production system following an exponential probability density function with different parameters. Tables 3.14, 3.15, 3.16 and 3.17 describe the information represented by the place nodes, the colors used, the transition nodes and the arc expressions, respectively.

Table 3.18 gives the production times in every one of the machines for each type of piece.

The initial conditions of place node P4 are:

Table 3.14 Description of place nodes of a system's CPN with 3 production sequences

Place	Color	Description
P1	C1	Stock of type $\Pi 1$, Π_2 and Π_3 pieces
P2	C2	Machine in *free* state
P3	C3	Machine in *working* state
P4	C4	Sequence de scheduled procedures
P5	C1	Stock of finished type Π_1, Π_2, and Π_3 pieces

Table 3.15 Definition of colors C, C1, C2, C3 and C4

Color	Definition	Description
CP	Integer 1…3	Identification of the type of piece
CT	Integer	Time assigned to a production operation
C1	Product CP*C2*CT	Identification of the type of piece, stock where it is to be found, as well as the time required in the following production operation
C2	Integer 1…6	Identification of machine in *free* state
C3	Product CP*C2*CT	Identification of type of piece, machine in *working* state and length of time of the procedure
C4	Product CP*C2*C2*CT	For each type of piece it specifies the completed procedure, the next production operation to be performed and the required time

Table 3.16 Description of transition nodes of a system's CPN with 3 production sequences

Transition	Description
T1	Arrival of type Π_1 piece
T2	Arrival of type Π_2 piece
T3	Arrival of type Π_3 piece
T4	Start production operation
T5	End production operation
T6	End sequence of procedures

Table 3.17 CPN arc expressions for a system with 3 production sequences

Arc	Expression
a1	1′(1, 1, 125)
a2	1′(2, 5, 105)
a3	1′(3, 1, 135)
a4, a6, a7	1′(x, y, z)
a5, a8	1′(y)
a9	1′(x, y1, z1)
a10, a11	1′(x, y, y1, z1)
a12	1′(x, 7)
a13	1′(x)

Table 3.18 Production times

Type of piece	Sequence procedures with times
Π_1	Adjustment (125′)—Pressing (35′)—Drilling (20′)—Polishing (60′)
Π_2	Molding (105′)—Milling (90′)—Drilling (65′)
Π_3	Adjustment (135′)—Molding (250′)—Milling (50′)—Pressing (30′)—Polishing (25′)

$$\begin{aligned} \text{P4} &= 1'(1,1,3,35) + 1'(1,3,2,20) + 1'(1,2,6,60) \\ &\quad + 1'(1,6,7,0) + 1'(2,5,4,90) + 1'(2,4,2,65) \\ &\quad + 1'(2,2,7,0) + 1'(3,1,5,250) + 1'(3,5,4,50) \\ &\quad + 1'(3,4,3,30) + 1'(3,3,6,25) + 1'(3,6,7,0) \end{aligned}$$

Chapter 4 explains how PN and CPN implemented in SIMIO.

References

CPN Group, University of Aarhaus. (2015). http://cs.au.dk/cpnets/

Guasch, A., Piera, M. A., Casanovas, J., & Figueras, J. (2003). *Modelado y simulación: aplicación a procesos logísticos de fabricación y servicios* (2a ed.). Barcelona: Ediciones UPC.

Jensen, K. (1997). *Coloured petri nets: Basics concepts, analysis methods and practical use* (Vol. 1, 2, 3). Berlin: Springer.

Jensen, K., & Kristensen, L. M. (2009). *Coloured petri nets: Modelling and validation of concurrent systems*. Berlin: Springer.

Kristensen, L. M., & Christensen, S. (2004). Implementing coloured petri nets using a functional programming language. In *Higher-Order and Symbolic Computation* (Vol. 17, pp. 207–243). Netherlands: Kluwer Academic Publishers.

Narciso, M. E., & Piera, M. A. (2015). Robust gate assignment procedures from an airport management perspective. *Omega The International Journal of Management Science, 50*, 82–95.

Narciso, M. E., Piera, M. A., & Guasch, A. (2010). A methodology for solving logistic optimization problems through simulation. In *SIMULATION: Transactions of the Society for Modeling and Simulation International* (Vol. 86(5–6), pp. 369–389).

Narciso, M. E., Piera, M. A., & Guasch, A. (2012). A time stamp reduction method for state space exploration using colored Petri nets. In *SIMULATION: Transactions of the Society for Modeling and Simulation International* (Vol 88(5), pp. 592–616).

Petri Nets World, Universidad de Hamburgo. (2015). http://www.informatik.uni-hamburg.de/TGI/PetriNets/

Piera Eroles, M. A., Narciso Farias, M. E., & Buil Giné, R. (2009). Flexible manufacturing systems. In *Simulation-Based Case Studies in Logistics. Education and Applied Research*. London: Springer.

Proth, J.-M., & Xie, X. (1996). *Petri nets: A tool for design and management of manufacturing systems*. Inglaterra: Wiley.

Silva, M., & Valette, R. (1989). Petri nets and flexible manufacturing. In *Lecture Notes in Computer Science, Advances in Petri Nets* (Vol. 424, pp. 374–417).

Wang, J. (1998). *Timed petri nets: Theory and application*. The Kluwer International Series on Discrete Event Dynamic Systems. Norwell, Massachusetts: Kluwer Academic Publishers.

Zaremba, M. B., & Prasad, B. (1994). Modern manufacturing: Information control and technology. In *Advanced Manufacturing Series*. Berlin: Springer.

Zhou, M., & Venkatesh, K. (1999). Simulation and control of flexible manufacturing systems: A petri net approach. In *Series in Intelligent Control and Intelligent Automation* (Vol. 6). Singapore, New Jersey, London, Hong Kong: World Scientific Publishing.

Zimmerman, A. (1995). Modeling of manufacturing systems and production routes using colored petri nets. In *Proceedings of the 3rd IASTED International Conference on Robotics and Manufacturing* (pp. 380–383).

Zimmermann, A., Dalkowski, K., & Hommel, G. (1996). A case study in modelling and performance evaluation of manufacturing systems using colored petri nets. In *Proceedings of the 8th European Simulation Symposium, ESS '96* (pp 282–286).

Chapter 4
The Coupling of Coloured Petri Nets with SIMIO

Miguel Mujica Mota

4.1 Introduction

This chapter is probably the one that makes the book different from others focused in Petri nets or Coloured Petri nets, since it introduces concepts and the methodology of how to integrate the Coloured Petri Nets (CPN) formalism with a discrete event-based simulation program. SIMIO simulation software was chosen as the appropriate software because of the following characteristics:

- It is a young software that uses advanced modelling approaches unavailable with traditional software, like *ARENA* or Promodel.
- It does not require knowledge of a particular programming language to set the particular functionalities of the model being developed.
- It has a very powerful graphic interface, which makes the models developed in CPN easy to understand for people who are not familiar with modelling and simulation.
- Integrating CPN with software like SIMIO makes allow their features to complement each other sufficiently so the modeller ends up with a very powerful system analysis and simulation tool.

SIMIO has been produced by the original developers of the ARENA simulation software. This software was created with a multi-paradigm modelling tool that supports both an object orientation and process orientation. In other words, all the objects used in the SIMIO work area (facility view) are, in essence, objects, but all the logic that is inherent to the objects is controlled by using processes that govern their behaviour. These processes are encapsulated in the objects, but new processes can be used for extending the logic of their behaviour (Pegden 2007; Kelton et al. 2010). The way how to integrate Petri nets with the modelling environment has recently been studied (Mujica and Piera 2011) since we identified, on the one hand,

I.F. De La Mota et al., *Robust Modelling and Simulation*,
DOI 10.1007/978-3-319-53321-6_4

that modelling environments are not good enough at helping the analyst to understand the cause-effect relationships in the systems, and on the other, the formalism of the Petri nets is too abstract to be presented to decision makers, who are most often the end users.

4.2 Review of the Methodology

As it has been mentioned, the approach presented in this book is a combination of CPN with SIMIO, the idea is to illustrate a beginner modeller or the user of either one tool or both tools how their models will be much more powerful with the combination of both approaches. As I emphasized previously, the user can take advantage of both techniques and analysis tools for making more robust models. In the case of CPN models we suggest the reader that, after developing the CPN model, he verifies the performance and correctness of it with a computer tool. I particularly suggest the use of CPNTools (CPNtools 2016) which was developed originally by the University of Aarhus in Denmark (Aarhus University 2016) under the supervision of professor Dr. Kurt Jensen who I had the pleasure to spent some time working in his group as an internship during my Ph.D. Currently the management of CPNTools has migrated to the Technical University of Eindhoven under the supervision of Dr. Michael Westergaard (2016). The use of CPNTools is relatively easy and all the different performance tests (such as fairness, boundedness, reachability among others) can be done in a straightforward way using the *State space analysis tool*. For a deeper understanding of the performance tests, I refer the reader to check the book of Jensen and Kristensen (2009) since the explanation of them is out of the scope of this book and there are several books which make an explanation of those tests.

After the model has been coded in CPN and its correctness verified with CPNTools the integration with a simulation program such as SIMIO can be performed. With the integration of CPN models in SIMIO we can include and evaluate other characteristics of the system under study such as variability of the system, several elements left out of the scope of the CPN models or simply run hundreds of experiments to understand the performance of the system at different levels of detail. Figure 4.1 shows the methodology for building powerful models using CPN and SIMIO.

4.3 SIMIO: Modelling Environment

As mentioned above, SIMIO combines simple object-oriented approach with the processes paradigm, allowing a high degree of flexibility in the development of models. This is particularly suitable for implementing the semantics of Petri nets.

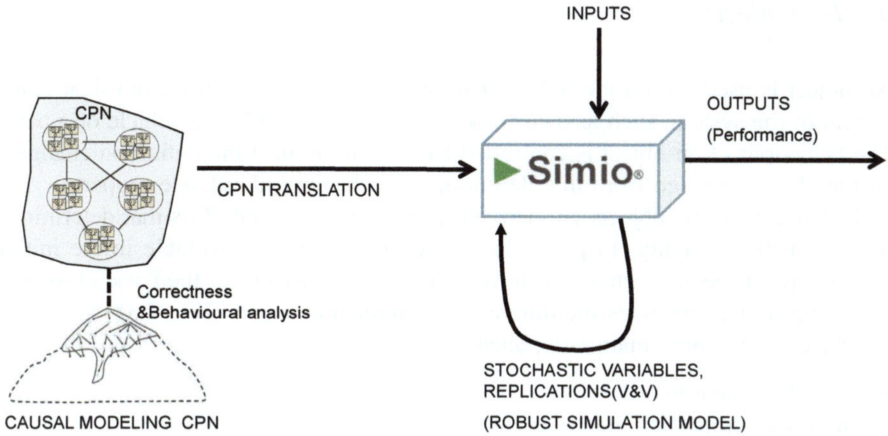

Fig. 4.1 CPN-SIMIO methodology

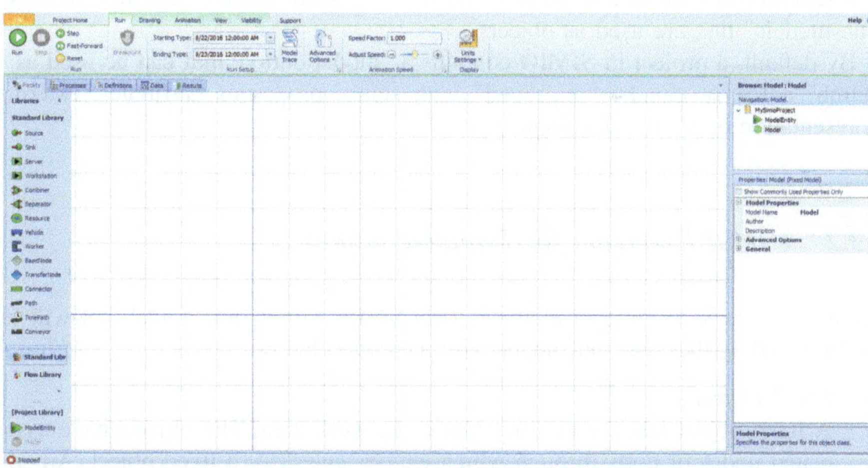

Fig. 4.2 SIMIO facility view

SIMIO (in its complete version) is composed of 6 sections: the facility view or work area, which is where the models are developed; the processes area, used to extend the functionalities of the objects; the definitions area, where variables, properties and other elements are added; the data area that is used to define properties by making use of tables; the dashboard that serves to monitor the evolution of objects in SIMIO; and the results area, where the results can be analysed. Figure 4.2 illustrates SIMIO's start window.

4.3.1 Objects

An object is used within the SIMIO Facility view and represents a physical component of the system, such as a customer, a worker, a machine, a vehicle or a path. The behaviour of an object is controlled by its definition; if the definition changes, all the objects derived from that definition will change (inheritance property).

In SIMIO, all the objects are derived from a more general class that determines their main functionality. In particular, the different objects available in the initial library have a specific behaviour derived from the behaviour of the basic class; the following figure illustrates the different hierarchical concepts (Fig. 4.3).

Objects have three main components:

- Model (behaviour)
- Interface (properties, states, events)
- External representation (entry/exit nodes, graphics).

Any model can be transformed into an object if an interface and an external representation are added. A project can have a "main model" and one or more "sub-models" that are used as objects.

By default, a project in SIMIO contains a "model entity" that can be used as a "dumb" entity or can be "improved" with states, properties, attributes, external representation or a particular logic.

4.3.2 Useful Elements for Implementation

Some of the most significant elements of a SIMIO simulation model that will be used for coding the Petri net models are mentioned below.

(a) *Fixed Objects*
A fixed object has a static location in the work area. The objects have associated nodes or ports through which entities enter/exit a fixed object. Among them we can mention the SOURCE, SERVER and SINK objects.

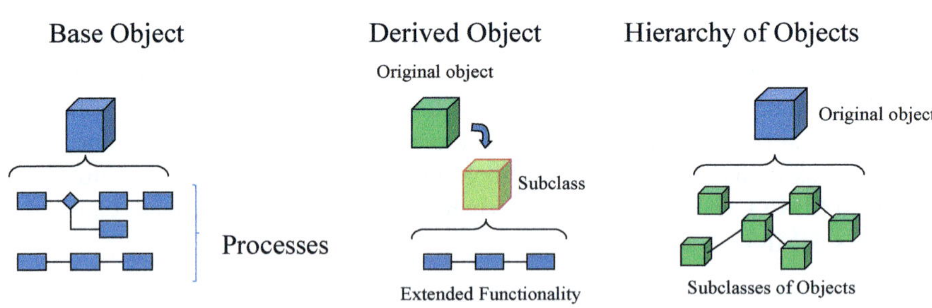

Fig. 4.3 Types and hierarchy of objects in SIMIO

(b) *Links*
Links define a path for entities and transport between two nodes, and can be unidirectional or bidirectional. Connections can be used to build networks. These links have a weight that can be used for selecting a path, based on each link's weight.

(c) *Nodes*
Nodes define the start and end point for one or more links. They are also used to model the intersection of multiple entry/exit links. The nodes can define the entry and exit points of an associated fixed object.

(d) *Entities*
Entities are objects that flow through the links, and enter and exit the objects; they can also belong to a particular network (determined by nodes and links) and follow a specific route sequence. Given that entities are objects in SIMIO, they can have logic that specifically controls their behaviour.
Generally, entities are generated by the SOURCE object, but can also be generated through the use of a process internal to the object and under specific conditions.

(e) *Stations*
Stations are elements that are found in the definitions area of SIMIO. The concept of station in SIMIO means a site where the entities that flow can be located inside the object. In this sense, and given that they are part of the anatomy of an object in SIMIO, they can be used to extend the functionalities of an object through inheritance (Weisfeld 2009) in order to develop a new object and be used independently in the model to store entities.

(f) *Attributes*
The concept of attributes is implemented in SIMIO through what are called *states*. These variables in SIMIO correspond to attributes already associated with the global model as the parent object or with the objects that form part of the global model which conceptually could be associated as local variables in any programming language. Their main characteristic is that the values they acquire will change as simulation proceeds.

(g) *Properties*
The concept of properties is coded in SIMIO through the so-called *Properties* of the object. These properties are parameters that define the object's morphology and the main goal is for these parameters to be defined during the process of developing the model. Because these properties are parameters that define the behaviour of the object, they can be added at your own discretion to get a more suitable model. Their main characteristic is that their values are set at the beginning of the simulation and will not change during the simulation run.

(h) *Tokens*
The concept of token is inherent to SIMIO, but the concept is different from its use in the semantics of Petri nets. This is important to mention as the reader might be somewhat confused on finding that SIMIO has a similar concept. The

concept of token in SIMIO serves to govern the Steps within the processes area (the logical execution of steps). We cannot see these tokens graphically but the different processes that define the logical function of an object are executed through the virtual flow of the tokens in the *steps* that comprise a process.

4.4 SIMIO/Petri Nets Equivalence

We cannot talk about a direct equivalence between either the ordinary or the coloured Petri nets (CPN) and SIMIO. However, it is highly useful to do the conceptual analysis with a modelling formalism like CPN and integrate the dynamics of the formalism into a program like SIMIO. Implementing this makes it possible take advantage of the features of both environments for a more robust simulation study.

Using SIMIO together with the modelling formalism lets us take advantage of the capabilities of both approaches. On the one hand, we can formally model the situations that cause the evolution of events in a system, on the other, we can employ the tools for analysis and graphic power present in a program like SIMIO. All this from a *Bottom-Up* analysis approach, i.e., we can construct the model progressively.

4.4.1 Equivalence Between the Dynamics of SIMIO and the Coloured Petri Nets

This section introduces the reader to a way of implementing elements from the coloured Petri net formalism to the SIMIO simulation environment. With the use of this section, the reader can directly implement models developed with the CPN formalism to SIMIO models in such a way that they keep the dynamic behaviour originally set out with the CPNs.

(a) *Tokens*
The entities that flow in SIMIO are used to achieve a correspondent relationship with the tokens of the Petri nets. When these entities do not have any attribute they correspond to ordinary Petri Net tokens, but they can correspond to tokens of the coloured Petri net formalism if we add attributes to them by using the SIMIO *states*. Figure 4.4 shows the correspondence of entities with tokens that have attributes (colours).
If we wanted to use tokens with COLOURS, we would have to define states for the entities of the SIMIO model. As shown in Fig. 4.4, these attributes can have different types of values: Integer, Boolean, String, etc.

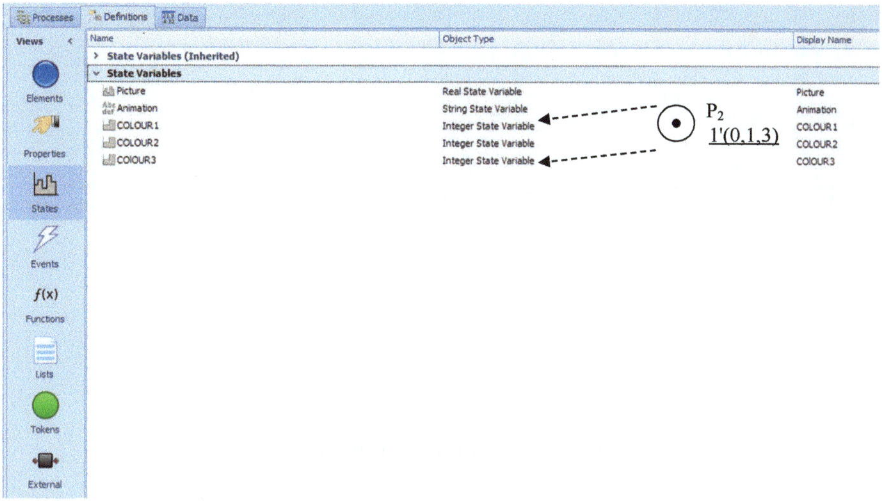

Fig. 4.4 Coloured tokens in SIMIO

(b) *Place nodes*
In order to implement models developed in Petri nets, the place nodes will correspond with the SIMIO *Stations*. The functionality of stations in SIMIO is to store the entities and make part of almost all the objects from the standard SIMIO library. Figure 4.5 illustrates the architecture of an object in SIMIO.

The stations are elements that can exist as independent objects or as objects that make part of other, more complex objects in the environment. If the stations are used

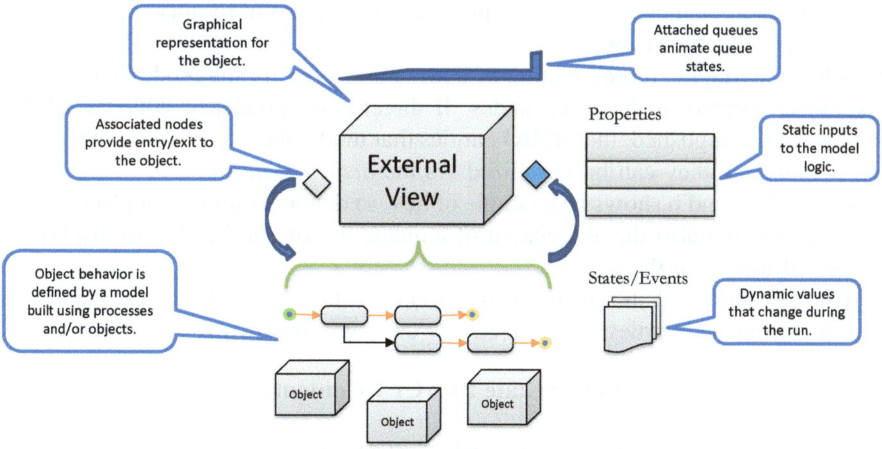

Fig. 4.5 Anatomy of an object in SIMIO (taken from SIMIO material)

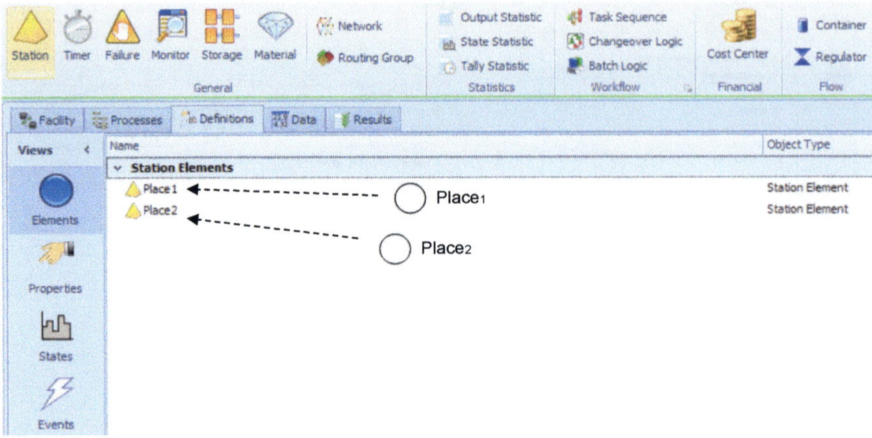

(a) Definition of Stations in SIMIO

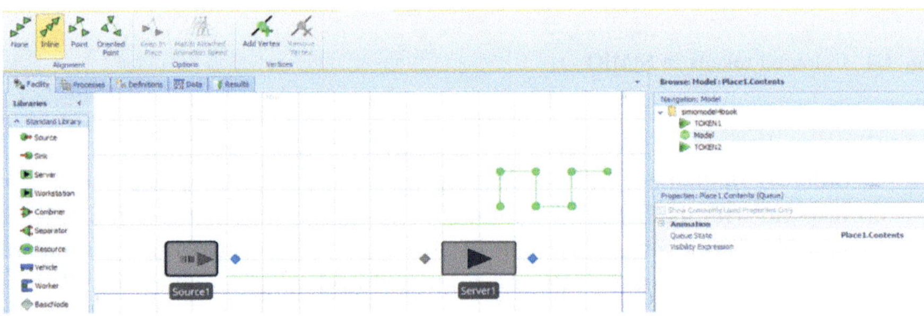

(b) Definition of stations in SIMIO

Fig. 4.6 Definition of stations in SIMIO

as independent elements, graphic display can be carried out by means of *detached queues* in the facility view.

When the graphic representation of Petri nets is used for the modelling task, the tokens are located in the place nodes. If direct correspondence with the SIMIO stations is accomplished, the SIMIO entities that model the tokens will be located in the stations and they can be visualized via the *detached queue*.

Figure 4.6a and b shows an example of how to define a station as a place P1 of a CPN model, and also the association of a queue in order to be able to display the storage of tokens in the stations.

The association of the queue with the entities that are found in the stations is specified in the properties of the queue by using the instruction:

Queue State|PLACE1.Contents

In this example the place node is called *PLACE1*.

4.4.2 *Conditioned Events and Satisfaction of Constraints*

Conditioned events are all those where a set of conditions have to be met in order for the events to be carried out. When analysing a system, the analyst commonly finds the conditions that unleash events within it. The conditioned events are suitably modelled by the CPN through the use of constraints that are implemented as input nodes, guards, arc weights and expressions within the CPN model. Unlike ordinary Petri nets, conditioned events in the CPNs are specified not just by the weight of the arcs and the input place nodes of a transition, but also by the arc expressions and the guards associated with the transition.

The arc expressions impose constraints on the colours of the tokens participating in the assessment of conditions that enable the transition. Separately, the *Guards* are Boolean expressions that impose particular constraints associated with the transition and the values of the variables in the arcs; the restrictions associated with the guards have to be satisfied together with all the aforementioned constraints in order to activate the transition.

Figure 4.7 illustrates a conditioned event that is typical of CPN formalism.

In order to develop an equivalent model in SIMIO we have to use the following elements:

- *Stations*
- *Decide steps*

Stations can be used independently as element, or as stations pre-located in the SIMIO objects as shown in Fig. 4.8. If stations from the objects of the standard library are used, the functionalities implemented in the library objects can be used, as might be the case of a *seize*, *delay*, *release* process, whose functionality has already been modelled in a *Server* or *Workstation* object.

The same figure illustrates two ways of implementing a conditioned event. Figure 4.8a shows a process diagram that uses stations as independent objects. The properties window of the step that is highlighted (*Transfer*) refers to a station called Place1. This station was created earlier in the object definitions area (*Definitions*).

Figure 4.8b presents the same logic, but in this case the *Parking Station* of the Node3 *Transfer* object is used, which belongs to the standard library of SIMIO.

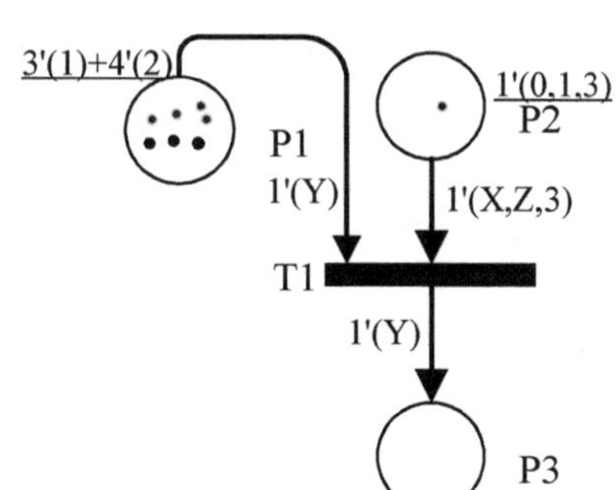

Fig. 4.7 Event conditioned by colours

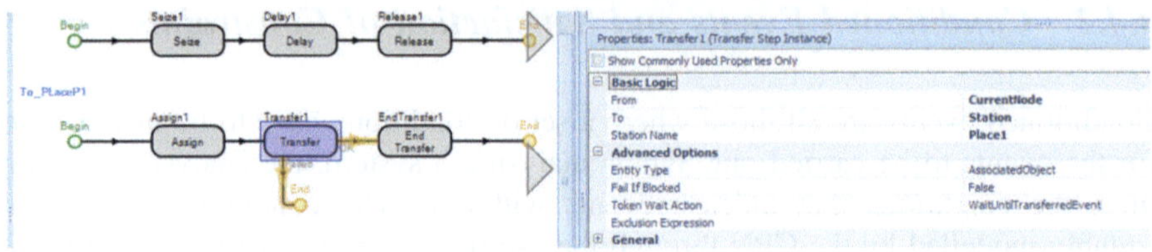

(a) CPN model

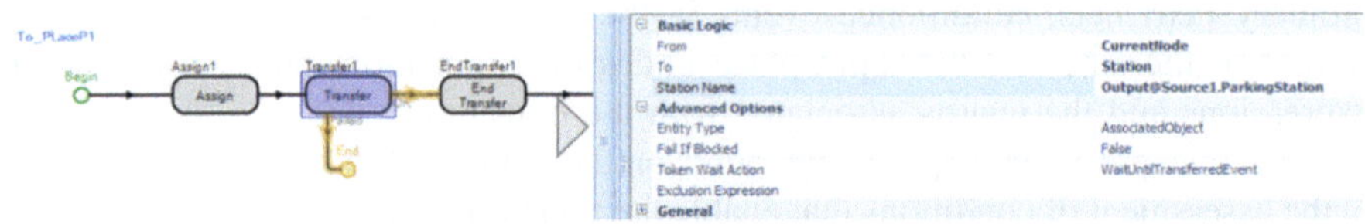

(b) Use of stations for storing entities

Fig. 4.8 **a** CPN model **b** Use of stations for storing entities

In this case it is up to the modeller whether to implement it taking the independent station or the one that belongs to the SIMIO object.

The evaluation of the constraints imposed by the arcs of a CPN model can be easily done by using the *Decide* step.

The *Decide* step is found in the *processes* area and its function is equivalent to that of an *IF.THEN.ELSE* of any programming language. The *DecideType* property must be specified as *condition based* in order to be able to define the expressions that must be satisfied. The *expression* attribute must contain the logic conditions that must be satisfied by the entity modelling the tokens of the CPNs.

Once the constraints have been satisfied, it is necessary to extract from the stations those entities that trigger the transition and generate the entities that correspond to the exit arcs of the CPN models.

Figure 4.9 shows how the satisfaction of constraints associated with the existence of enough tokens to set off the execution of the activity is verified.

On the one hand, and assuming we have a pair of stations called P1 and P2, a *decide* step is used to verify the existence of tokens in the two places, as observed in Fig. 4.9.

In this example, the associated expression (that must be coded in the properties window) for carrying out the verification is as follows:

$$\mathbf{P1.Contents > 0\&\&P2.contents > 0}$$

This expression is used to verify the existence of entities (playing the role of a token) in the two stations (playing the role of place nodes). If the arc expressions of the CPN model have specific information, such as constant values, this condition is specified making use of the *match* condition, which is an attribute of the *Search* step (see Fig. 4.10).

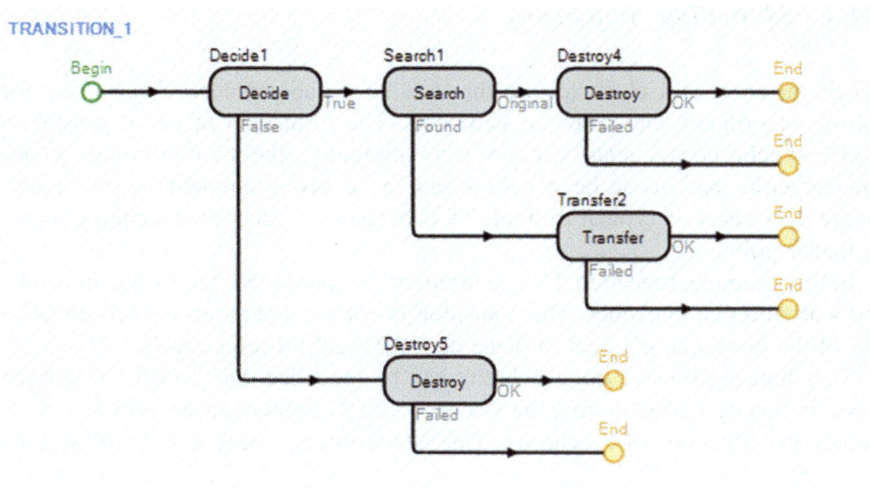

Fig. 4.9 Implementation of constraints

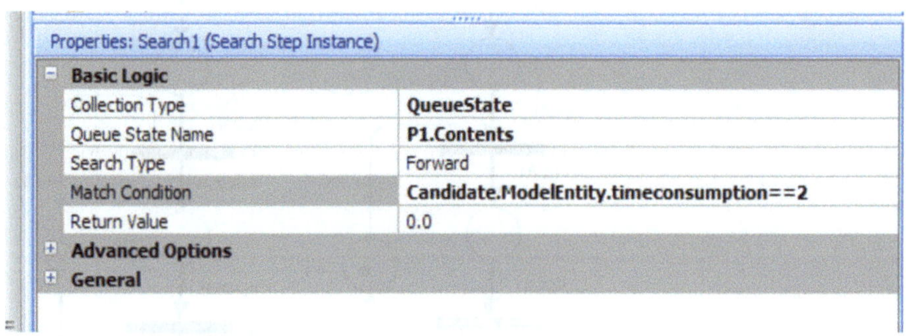

Fig. 4.10 Properties of the search step

The match condition lets us establish a search based on a characteristic and thus specify constraints on the search in order to find entities in the station that satisfy this condition. In the example of Fig. 4.10, the search is forward and done on the queue of the P1 object and the specified attribute must be fulfilled by the object being assessed (the *Candidate*). In this example, the property *timeconsumption* must be equal to 2 units. If entities are found that satisfy the specified characteristics, the logic flow continues through the *found* exit of the *search* step (Fig. 4.9) and from that step, the found entities are sent (using a *transfer*) to the location inside the *Facility*, where they will be inserted inside the *Facility* area of the SIMIO model.

4.4.3 Modelling Synchrony

Synchronization is a characteristic that can be suitably modelled using the formalism of ordinary and coloured Petri nets. The formalism makes it possible to model synchronous events without any ambiguity; the two processes evolve independently and must be synchronized, at a given moment, to do a task. Figure 4.11 shows a typical example of two processes that are synchronized at a particular point.

In this example, transition T5 synchronizes two processes that evolve in parallel and with different dynamics. This transition is not triggered until the token of each one of the processes is found in place nodes P4 and P8 respectively.

The characteristics of the synchrony can be modelled with SIMIO in different ways. In this case objects from the standard SIMIO library can be used that model already the situations of synchrony. This is possible by using a *Server* object and

Fig. 4.11 Synchronous petri nets

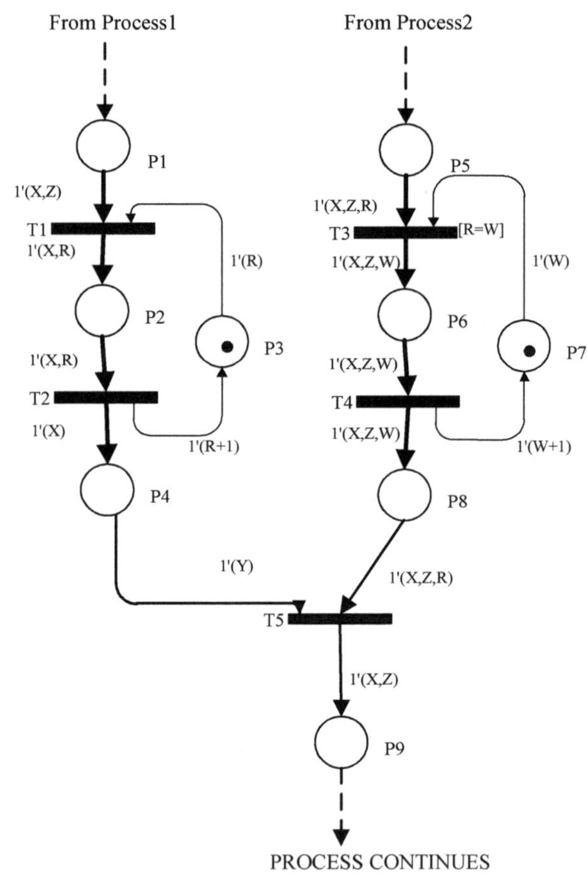

the *Secondary Resources* property. This property will be used to specify that the process cannot start until the resources specified by the property are available.

On the other hand, synchrony situations by adapting the CPN semantics can be simply modelled, by making use of the following elements:

- *Decide step*
- *Station*

In this case, once again, the stations can be used as independent elements or as objects of SIMIO from the standard library.

In the case of synchrony, it is possible to model the implementation of synchronal conditions by using procedures where conditions are established following the aforementioned steps, as given in Fig. 4.12.

In this example we use the expression of the *Decide* step to check if the conditions for starting an activity are fulfilled, in the example, two place nodes are defined making use of a pair of stations (P4 and P8). If the above procedure is inserted in an event associated with the procedure used to model operational flow, the *Decide* step checks whether there are entities or not in each one of the places where we want to model the synchrony. If it is not the case, then the original process will not continue. This logic could be used for modelling the synchrony of the place nodes that were mentioned before (place P4 and P8 of Fig. 4.11).

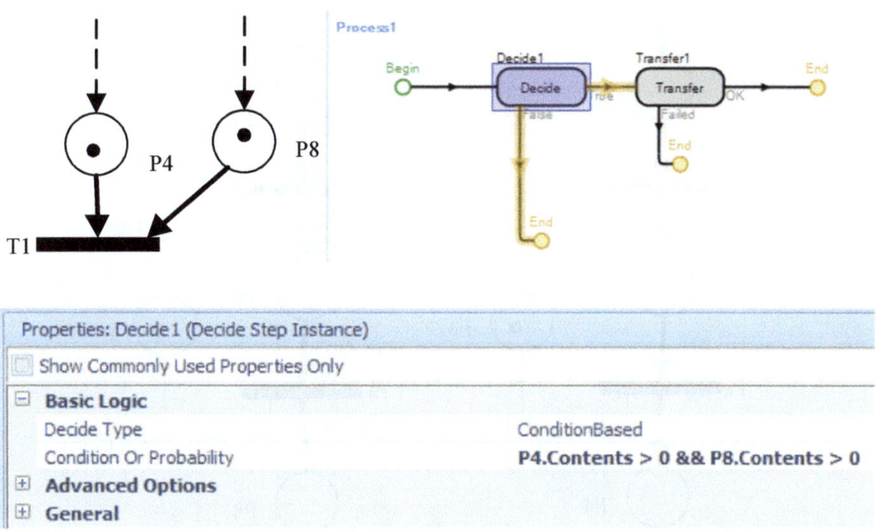

Fig. 4.12 Modelling in SIMIO the synchrony between two independent processes

4.4.4 Modelling Parallelism

Parallelism is the occurrence of processes that evolve from an activity or transition in common. These processes can evolve simultaneously or with different dynamics, beginning with the initiating transition; there may also be elements common to both processes. A typical case may be the modelling of flexible manufacturing systems. Figure 4.13 illustrates a CPN model with a dynamic like that.

The dynamics of parallel evolution of activities is simple to model. Figure 4.14 illustrates the implementation of two processes that will be triggered from one common event in the SIMIO process window. The steps participating in the implementation are the following:

- *Stations*
- *Set Node*
- *Decide*
- *Create*

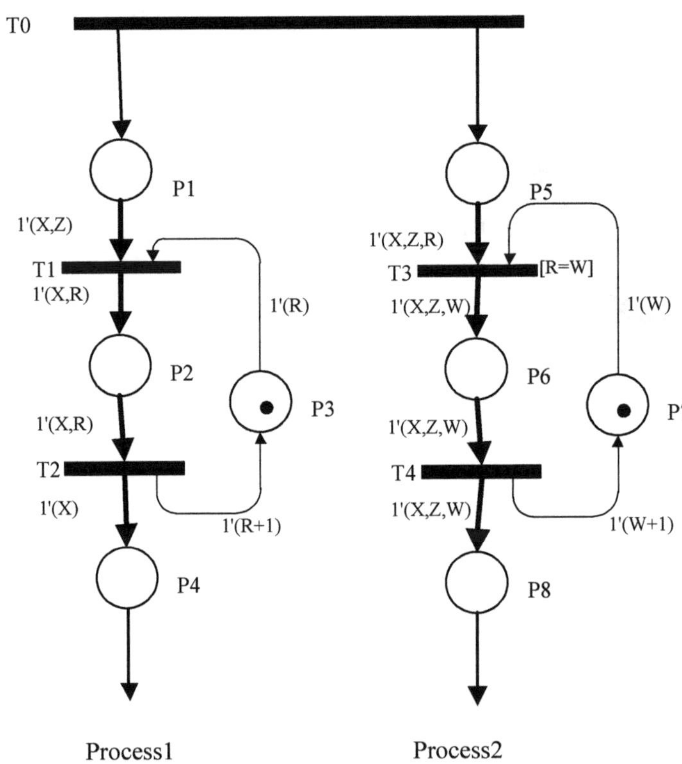

Fig. 4.13 Parallel processes in CPN

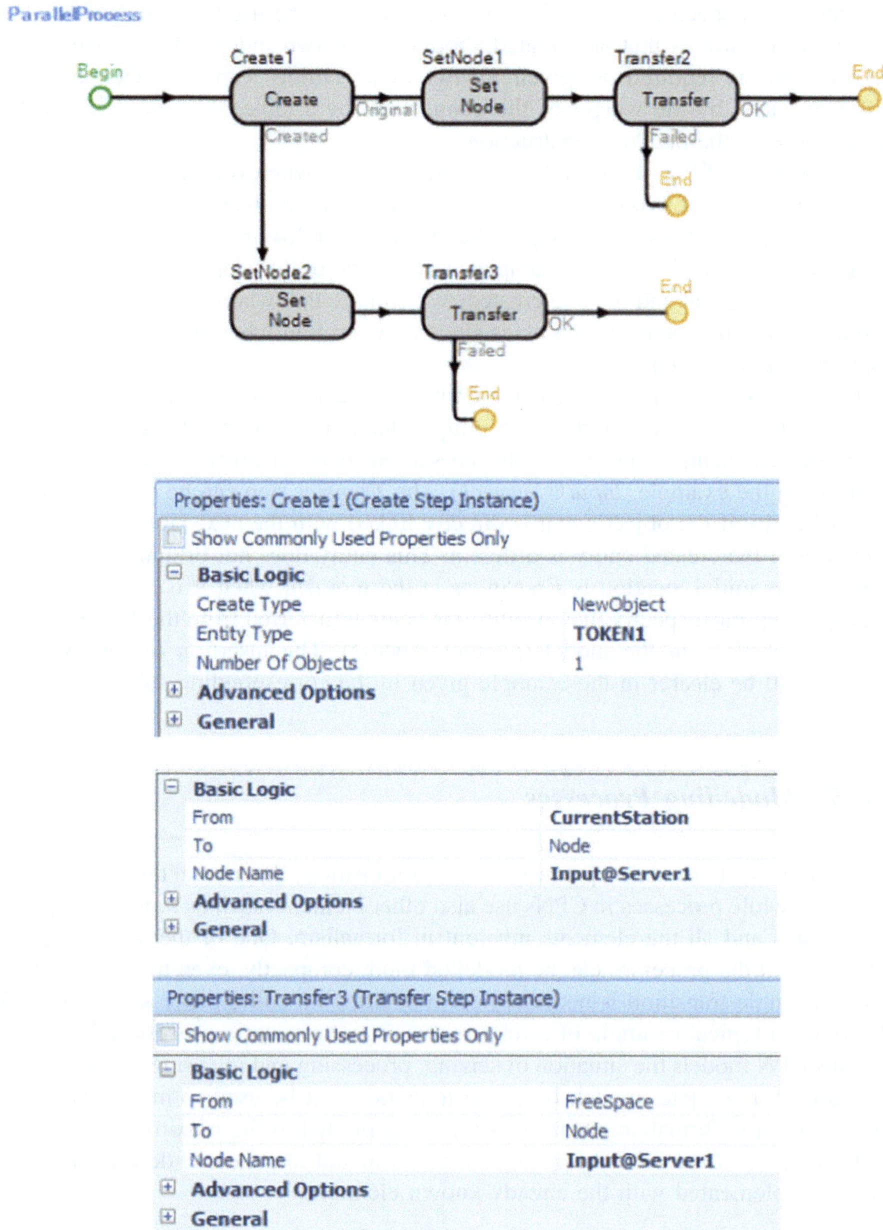

Fig. 4.14 Implementation of parallelism with SIMIO

Again, a simple implementation is possible, although in this case a pair of new steps must be introduced. On the one hand, the *Create* step will be used so that two parallel processes with independent evolution can be started using one entity that originates in a specific process. The *SetNode* step will be used to stipulate in what direction the entities that are created should go for two independent processes to evolve. This is required as, given SIMIO logic, entities can be routed in three different ways: by the weight of the connectors, by a table of sequences, and by making use of the *Set Node* instruction.

The process of Fig. 4.14 can be implemented from wherever we want to start the parallel evolution of two processes. As it can be observed, the *Create* step is used to create a new object (in this example, the entity called *Token1*). It is only necessary to create a single entity for the example, as at the end of this step there would be two entities, the original and the one created. We can see that when they both leave, the entities follow independent processes since they are told where to go in the model using the *SetNode* step.

Lastly, with the *Transfer* steps the entities are sent to their point of insertion in the SIMIO model. It is important to highlight that in the example, the original entity is transferred (using Transfer2) from the station, node or object to a specific destination (in the example, *Input@Server1*). The *Transfer* step can be omitted for the original entity if the object's exit is directly linked with the next process or object. The case of the created entity is different. This entity does not originally exist and therefore its initial location is *FreeSpace* at the moment when it is created; so the *Transfer* step must specify that the entity is being transferred from the *FreeSpace* to a particular place in the model (parallel process). The insertion of this type of process will be clearer in the example given in the corresponding subsection.

4.4.5 Modelling Processes

Using ordinary Petri nets, a process is a set of transitions that model the evolution of a system, while processes in CPNs use also other elements such as transitions, place nodes, arcs and all the elements inherent in formalism. One of the advantages of CPNs is that the processes can be modelled more compactly, even to the point that just one single transition is necessary for modelling an entire process. Figure 4.15 illustrates a typical example of a resource used to carry out a process (P3).

This CPN models the situation of seizing, processing and releasing a resource. In the case of a compact model, only one transition can be used to model the same process but the formalism characteristics are exploited to the maximum.

In the case of the SIMIO implementation, a model like the one described above can be implemented with the already-known elements.

- *Stations*
- *Set Node*
- *Decide*

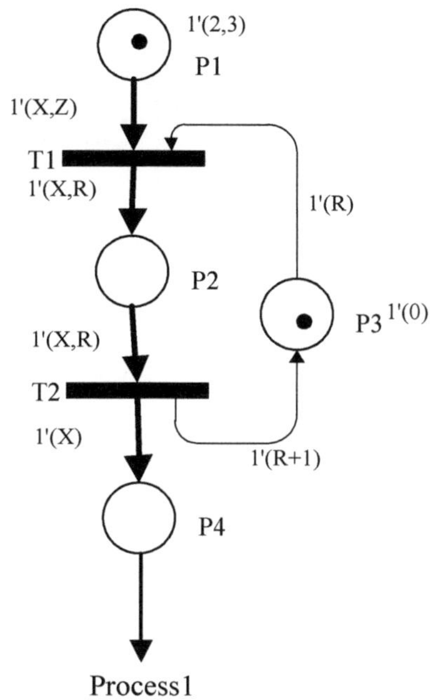

Fig. 4.15 Modelling of CPN processes

The seize, temporary delay and release of resources will also be used:

- *Seize*
- *Delay*
- *Release*

As it is to be expected, besides being able to define the simulation by using steps, there are some objects like the *Server* whose internal logic already has the aforementioned logic. Figure 4.16 shows the implementation of the steps for modelling resource seizing.

The *Seize* step enables us to specify which object will be the resource used for this process. During the resources-seized period, it cannot be used by another process. The *Seize* step makes it possible not just to seize a single resource, but to seize two or more resources. The example of a single resource would be equivalent to the implementation in Fig. 4.16. Using the *Delay* step, the time the modelling operation will take is specified. This operation can be carried out in a constant or random time; this time is specified in the *Delay Time* field in the properties window corresponding to the *Delay*. Lastly, the *Release* step is used to specify that the process has ended and therefore, that the resources used can be released.

The above sequence of steps can be inserted into any SIMIO object (nodes, objects, links among others) associated with some event of the objects. Furthermore, as i mentioned earlier, there are objects like the *Server* and *WorkStation* whose native logic has already implemented these sequence of resource seizing and release steps. Consequently, to use them, we only need to add

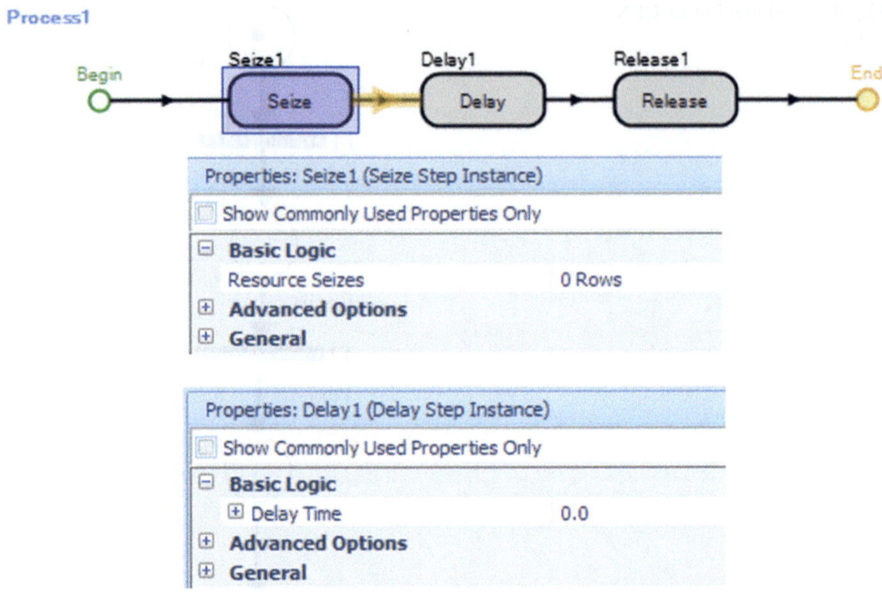

Fig. 4.16 Modelling a process with resources seize

the mentioned objects in the *Facility View* and to specify in the properties of the object the resource(s) that is/are required to accomplish the operation.

4.4.6 Modelling Queues

The modelling of queues is important in every dynamic system. Using the CPN modelling formalism, queues can be easily modelled with the use of a place node where the tokens present equal the number of elements in the queue of the real system. Figure 4.17 illustrates how this situation can be modelled.

The model might be representing a typical case of a queue at a bank or in a system of services where the customers are served on a FIFO policy. Transition T0 is always active and therefore can, at any time, generate an entity in place node P1. In this case, place node P1 represents the queue of entities or people who are waiting to be served by a resource (which could very well be a person) that is modelled by the token inside place node P2. With this representation, the queue would be symbolized by the quantity of tokens to be found in place node P1.

Queue modelling is easy with SIMIO. Generally speaking we can say that in practically all the predefined SIMIO objects that have a *Station* it is easy to display the queue of entities inside the station. When the objects from the standard SIMIO library are used, we can see the majority of them have associated green lines. These lines will move together with the object when moved by the user in the Facility

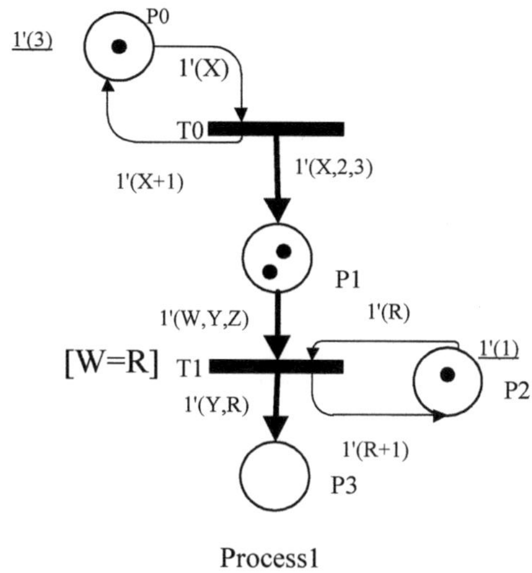

Fig. 4.17 Modelling queues in CPN

View. Figure 4.18 illustrates the location of the lines that allow us to visualize the queues for some objects.

The figure shows the lines in question which are used to display the queues that are formed in the corresponding station. The tables below demonstrate that when one of them is selected, its properties show the expression that associates the *Processing* station with the corresponding queue. Clearly, if a station is defined anywhere in the *Facility View*, the entities can be displayed by associating a queue (*Detached Queue*) with the station in which the expression will be similar to the one presented in Fig. 4.18.

4.4.7 Shared Resources

In every modelling formalism, it is extremely important to model resources properly; this is because different capacity indicators and levels of the resources present in the system can be determined through the use of simulation. In the case of CPNs, resource modelling takes place as described in the section that discusses processes. Figure 4.19 gives an example where we have two processes that share resources with different characteristics.

In the case of SIMIO, as described in process modelling, the use of the *seize*, *delay* and *release* steps can specify that a particular resource is required and this resource can, of course, be the same for both processes. It is important to point out that all SIMIO objects are resources by definition (Pegden 2007).

Thus, if the *Seize* step specifies the seizure of an object, this object will be disabled while it is being used.

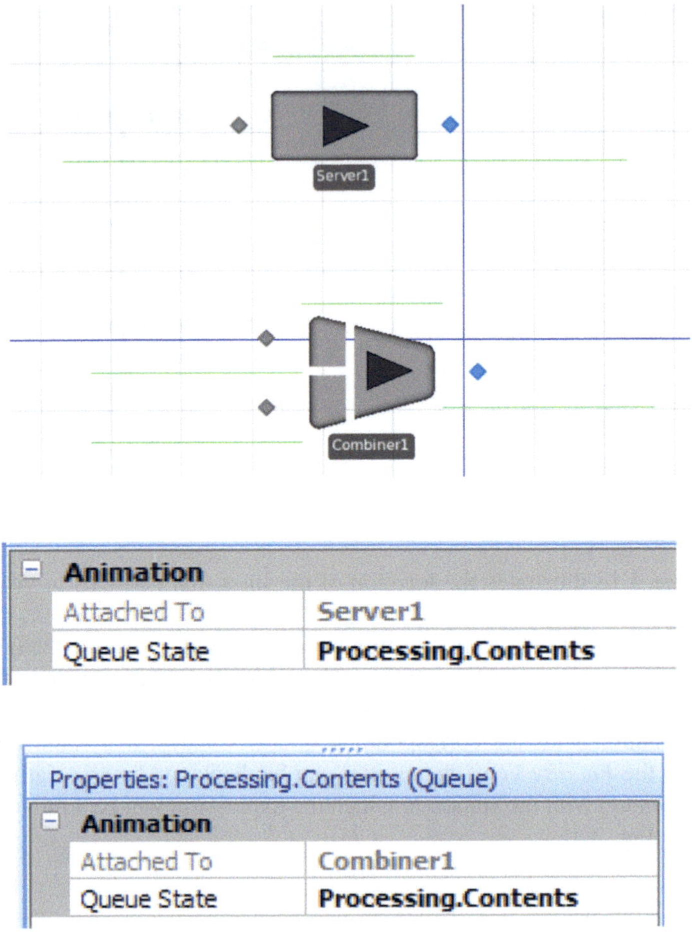

Fig. 4.18 Graphic implementation of the queue in SIMIO

4.4.8 Time Consumption

Time consumption is crucial for the ability to simulate and analyse how systems function over time. Using Petri nets, time consumption happens every time a transition is triggered (given that they model an activity). In this case, the trigger of a transition will be performed once the constraints imposed by the arcs, the arc expressions and the guards (in the CPNs) are satisfied. When the transition is executed, the time that is specified for the transition in the CPN semantics will be consumed.

In the case of SIMIO models, time consumption is carried out by using the *delay* step once the conditions that model the transition have been satisfied; consequently this step is located just after the entities that satisfy the constraints are extracted in

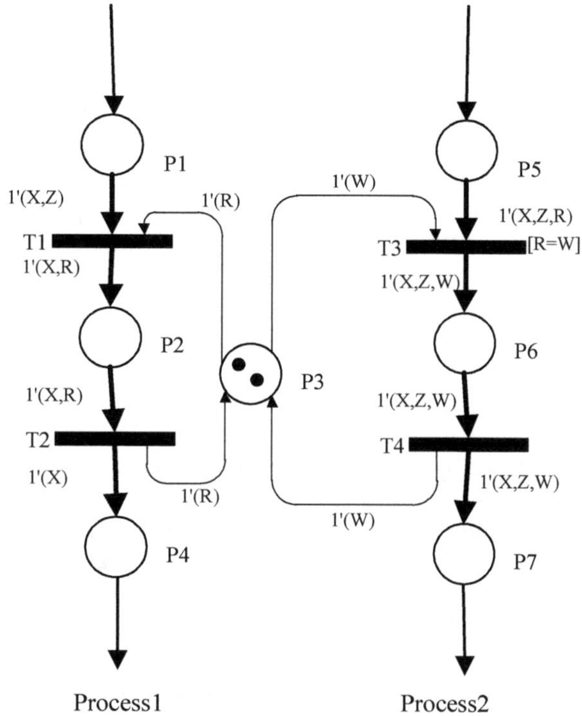

Fig. 4.19 Modelling the use of resources with CPN

the process window of SIMIO. One of the reasons for using CPN in a modelling setting like SIMIO is its possibility of taking advantage of the analysis, graphics and implementation characteristics of the environment in conjunction with the inherent traits of the modelling formalism. Apart from stations and SIMIO processes, it is a good approach to use objects such as the *server*, *path*, *or vehicle* that use already the logic of the *delay* step. This is easily done if, when the constraints are satisfied, the entities are sent to an object that does the time delay operation (performance of an activity). In addition, the time modelling using SIMIO can be more realistic as you can specify in the SIMIO objects (in the *delay* step) whether the time consumption is deterministic or stochastic.

4.4.9 Insertion of Petri Net Transitions in SIMIO

Due to the architecture of SIMIO, CPN rules associated with different events can be implemented in the logic of the objects. Each object has a particular number of events, which can be *EntityEnter*, *Processed*, *EntityExited*, among others. In this case, it is important to determine which SIMIO event will execute the logic associated with the CPN. As a general rule, given that what we want is for the CPN to "govern" the behaviour of the logic from a certain point in the simulation model; the CPN logic will be inserted at a point where the entity's characteristics are not affected by the inherent logic of the object where the entity is residing. So it is suggested to implement the CPN logic in the event associated with the entity entering to the SIMIO object.

Once the entity enters the object, its behaviour is governed by the semantics of CPNs and the next action is to specify the point where the entities re-enter the model in SIMIO.

4.4.9.1 Process Activators

The SIMIO simulation environment has the so-called Add-On Process Triggers, which are used to extend the functionality of the original object with processes developed by the user. Figure 4.20 illustrates the window of properties where these activators are found.

The model's properties window shows that there are different properties that can be modified, such as *process logic*, *Buffer capacity*, etc. These properties will vary depending on the object involved, but essentially all the objects will have these activators. By expanding the selection of *Add-On process triggers*, some events associated with the object drop down. It is precisely in these events where the association of the CPN model with the Simio one will take place.

The CPN logic can be integrated with the modelling environment in two ways. The first is called automatic link. This way of implementing the model in the

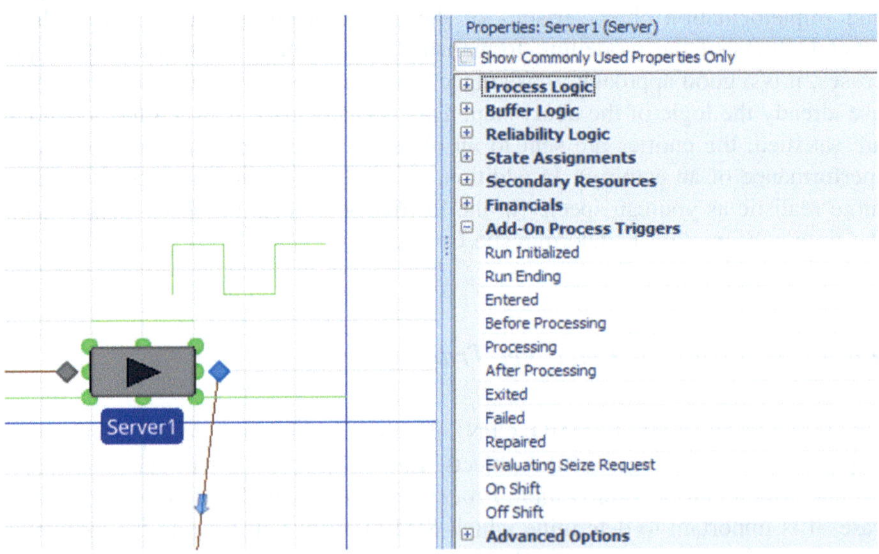

Fig. 4.20 Process activators

environment results from, first of all, selecting the event where we want to implement the logic of the transition to be assessed. We do this by selecting the event in question, e.g. *Exited* or *Entered*, and double-clicking on it. On the second click SIMIO will automatically generate the name of the process and send us automatically to the process area and placing the cursor in the process to construct.

The second way to implement the CPN is by manually linking, where the user first, as exemplified earlier, codes the CPN transition in question in the process area and later, makes the link to any event of any object from the environment. This particular manner of integration has the risk that, if the manual link is forgoten (because of carelessness on the part of the user), the process will not be activated in any way, because no event will exist that makes the call to the procedure, even though it has been coded in the process area. So we strongly suggest the reader to follow the first method.

4.5 Examples of Coloured Petri Net Implementation in SIMIO

We shall now give the reader an example of how to integrate a CPN model with SIMIO in order to use the analytical characteristics of both the simulation software and the formalism. For the sake of simplicity we present two simple models. The objective is that the user gets the clear idea of how to use this approach. The models might appear simple to the user but it does not mean that complex-relationships models cannot be coded; the author has developed complex models in which this approach is perfect for determining the causal relationships which could not be efficiently model with other approaches as we have emphasized in the previous sections.

The use of the methodology is illustrated with two models, one is a sub model which is part of a Boarding Model (Mujica and Flores 2015) in which the cabin of an aircraft is modelled; this model has been used to identify the causes of failure in the boarding policies in an aircraft. The second one is the model of a typical manufacturing system for which we should investigate the performance indicators and the best way of managing the shared resources which is key for improving any system.

4.5.1 Example 1: Boarding of Passengers at an Aircraft Cabin

The following CPN model is used for modelling the boarding process of passengers at a cabin of an airplane.

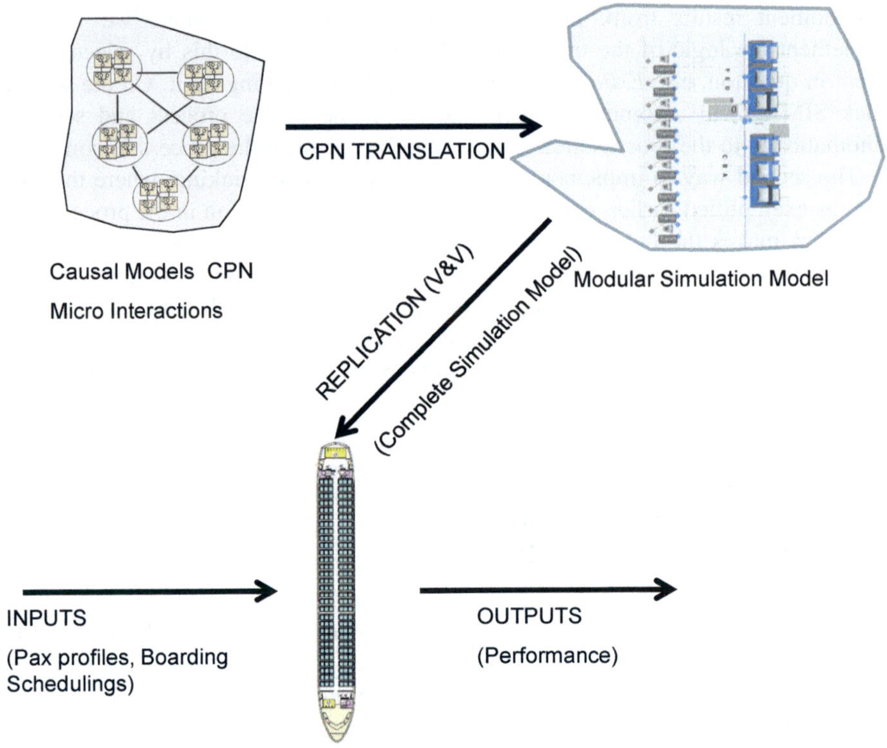

Fig. 4.21 Modelling approach for the use of CPN and SIMIO

In this example we present a bottom-up approach in which first we develop the logic behind the boarding processes at every row of a cabin. Then we implement the logic into one SIMIO model for performing a micro-simulation of the actions that happen at every row. Once we have verified that the behaviour is correct we are able to construct instances of rows connected through a common path in order to have a final version of the cabin of an aircraft. Figure 4.21 shows the modelling approach used in this work which is an extension of the methodology presented in Fig. 4.1.

As the diagram suggests, one first will develop the CPN model for modelling the micro interactions at the row within the cabin. Once the model is performed and verified using behaviour analysis for CPN (Jensen 1997) the next step is translating the logic into SIMIO in order to develop a model whose behaviour has been formalized and analysed using CPN. At this stage it is possible to validate and verify again the behaviour of the row at the cabin but taking into account other characteristics such as speed of passengers, age, blockings among others. When the SIMIO model is verified (modular simulation model in the figure), one can instantiate the number of rows of a cabin and connect them through a path in order to develop the complete behaviour within an aircraft cabin.

4.5.1.1 Coloured Petri Net Model

The CPN model is composed by 16 transitions and 3 place nodes. These transitions are the ones that model the different events performed at one row of the cabin. The formal definition of colours is presented in Table 4.1.

The model uses three place nodes for the modelling of the behaviour at every row in the cabin. The names assigned to the place nodes are called SEAT, AISLE and PAX. As the name suggests every place node is used for tokens which hold different type of information. The colour sets or multi sets, as they are known in the formal way, that are present in the different place nodes of the model hold different information are described in Table 4.2.

The following figures present examples of the different transitions that compose the model. As it has been mentioned, the model is composed by 16 transition nodes that represent the different activities or processes that happen when a person is arriving at the seat within a cabin. These activities could be like finding the seat completely empty or with people already sit and the actions that happen when people is sit or not. The information in the three place nodes will be used for modelling these situations.

The initial CPN model can be constructed and analysed using tools for CPN such as CPNTools (CPNTools 2016) which has behavioural analysis tools for verifying the correct behaviour of the model prior to its integration with the DES software.

Figure 4.22 illustrates one event when a passenger has to sit at the window (w = 100) and the row of seats is empty (y = 0). This could be the case either because of two situations, one is because nobody has sit yet (z = 0) or because there were passengers already sit but they had to stand up to let the passenger to sit in

Table 4.1 Colour definition and description

Colour	Definition	Description
X	Integer	It is the row number of the seat block
Y	{000, 001, 010, 100, 011, 101, 111}	It describes the seats occupied by the passengers. 000 means no passenger seated, 001 represents one passenger sit in the position closest to the aisle, 010 is used for representing a passenger sit in the middle and 100 represents a passenger sit in the window
Z	Integer	It represents the amount of people waiting in the aisle for the passenger to sit
R	Integer	It represents the row where the passenger is supposed to be sitting
W	{001, 010, 100}	It represents the seat location of the passenger. It is similar to Y
D	{0, 1}	It represents if the waiting person is seated in the middle (0 for either window or aisle and 1 for middle) and it is also used to represent that the passenger belongs to the right block (0) or the left block (1)

Table 4.2 Description of places

Place	Colour set	Description
SEAT	Product X*Y*D	This place represents the information of how the row is occupied. The first colour is used for the row number, the second for the seat occupancy and the last colour for the side of the block used
AISLE	Product Z*D*D	This place holds information about the amount of passengers standing up in the Aisle waiting for the seat, if they are sit in the middle, and which side they belong to. Z represents how many passengers are standing, the first D is used to mark if some passenger belongs to the middle, and the second D is to keep track of the side of the block the passenger belongs to
PAX	Product R*W*D	This place holds the information of the passenger. The first colour refers to the number of row, the second refers to the position of the passenger in the seat block (window, middle, and aisle) and the third one is the information about which side of the row he belongs to

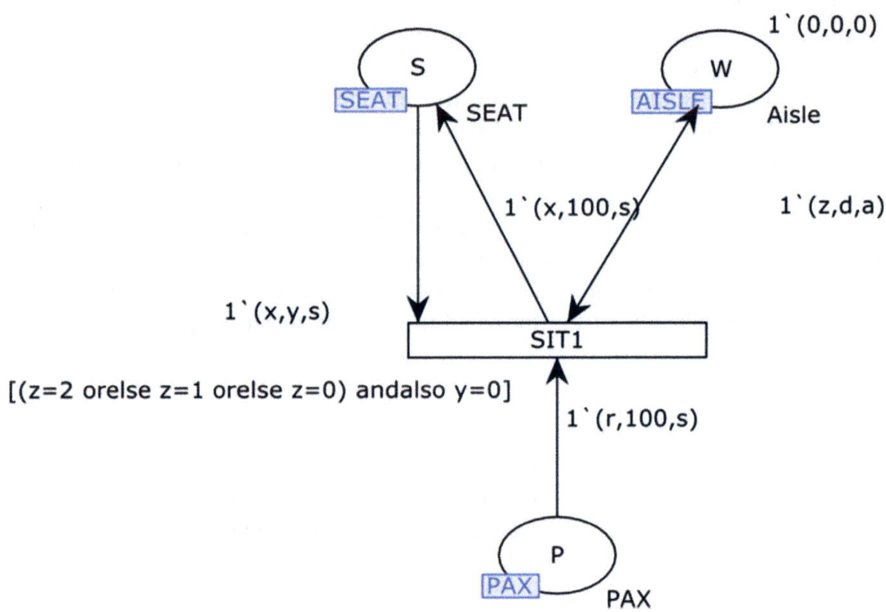

Fig. 4.22 Transition for sitting

Place W (z = 2, z = 0 or z = 1). Under this situation the corresponding time consumption can be associated to the transition, but it would depend on the correspondent study. Once the passenger is sit, the new colour value is assigned via the output arc to the place node S with the value of variable *w* which in this example is 100 [1'(x,100,s)] representing that the passenger has reached his seat.

Another example is illustrated by Fig. 4.23 which represents the situation where a passenger must get to his seat at the window (w = 100) and the middle seat is

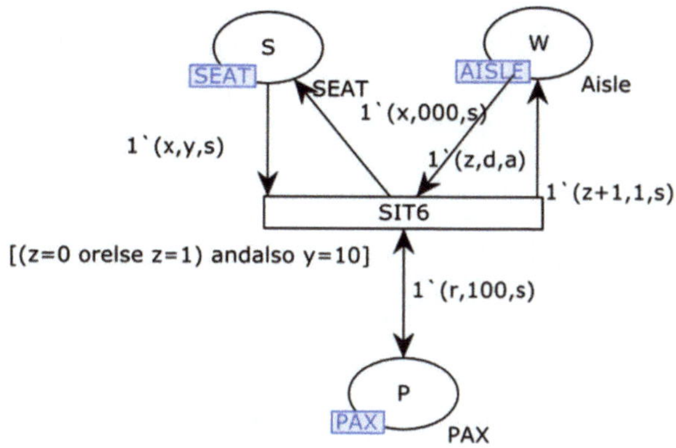

Fig. 4.23 Walking out of a passenger seated in the middle

occupied by another passenger who was previously sit (y = 10). The latter passenger must walk out so that the passenger at the window can reach to its seat. The following transition models the action of walking out by the passenger that was sit and then he must wait at the aisle for the other passenger to take his seat.

In this model a unit is added to the colour z (i.e. [z + 1]) of the token in the waiting place (Place node AISLE) and the variable *d* turns to 1 (1 represents that the passenger that stood up was from the middle) and the value *s* is used for keeping track of what side of the row the passenger belongs to. The token in the PAX place node do not change values since the event is that the passenger in the middle goes out. Finally with this event the token in the place node SEAT changes its *y* value to 000 to represent that the seat is now empty.

Using the CPN approach we formally specify the cause-effect relationships and when the model is simulated, emerging dynamics appear that sometimes hinder the smooth flow of passengers inside the cabin during the boarding or deboarding process. In addition, more colours can also be easily added to the model to represent characteristics such as age, size, number of bags, disabilities etc. and those characteristics can be used to simulate events in a more accurate way and then the emergent dynamics that appear once the model is developed are more accurate. The total model is composed by 16 transitions that represent all the events that appear during the seating of passengers at one row of the cabin. Figure 4.24 presents the different transitions of the CPN model.

4.5.1.2 Modular Integration

The CPN models previously developed are in turn integrated in SIMIO, a DES-based software tool, following the rules and implementations presented by the author in a previous work (Mujica and Piera 2011). The CPN model is developed in such a way that the symmetry present in the cabin is used for avoiding the re-coding of redundant logic. The cabin can be simulated using a module that represents one row and then the module (governed by the CPN) can be instantiated to develop a

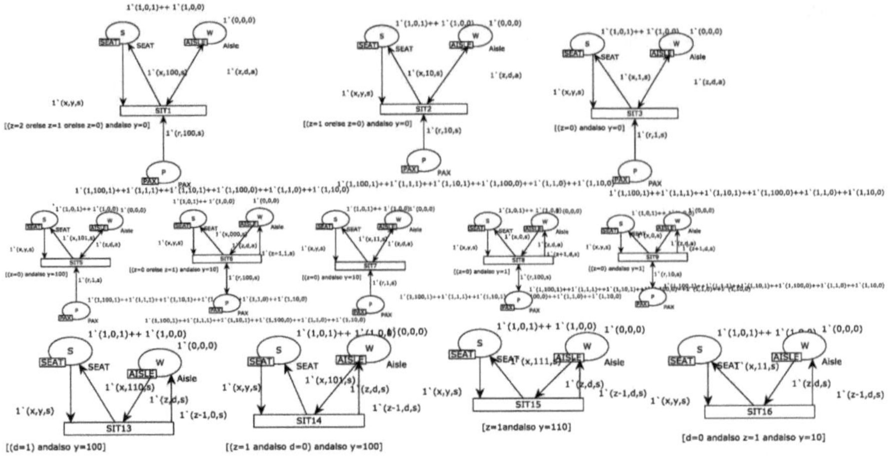

Fig. 4.24 Some transitions of the CPN model

complete model of the cabin. SIMIO is very efficient for this approach since it has an object oriented approach in which modularity is inherent in it, but the approach can be also implemented using another DES-based tool such as ARENA or ANYLOGIC.

The resulting simulator models with high accuracy the micro interactions between passengers, and the emergent dynamics are assessed when the total model is developed. It is fair to mention that the implementation allows also taking full advantage of the software capabilities and simulate accurately the stochasticity inherent in the system which otherwise could take long time to develop using the formalism alone.

The approach is the one presented in Fig. 4.21; first, it is necessary to implement the different activities (transitions) in a module that simulates one row of the cabin. Second, advantage is taken from the use of a modular approach when the different rows of the cabin are put together in serial order. With this action we are able to make a complete model for the cabin that takes into account not only the micro-interaction between passengers (at row level) but also the interaction that occurs at higher levels i.e. in the aisle, walking speeds, aisle blocking etc.

Figure 4.25 illustrates the elements for the row-module in which the CPN model steer the evolution of activities and events during simulation time. The methodology proposed by Mujica and Piera (2011) is used to implement the different transitions that occur during the seating process. The CPN logic of the transitions is coded in *Separator* objects from SIMIO in order to evaluate the different events that occur in the module. The transitions are evaluated concurrently using *Connectors,* which do not consume simulation time, so the logic associated to each object (CPN transitions) is evaluated all at once and only those that satisfy the different restrictions are fired thus the simulation is performed with high accuracy. In the

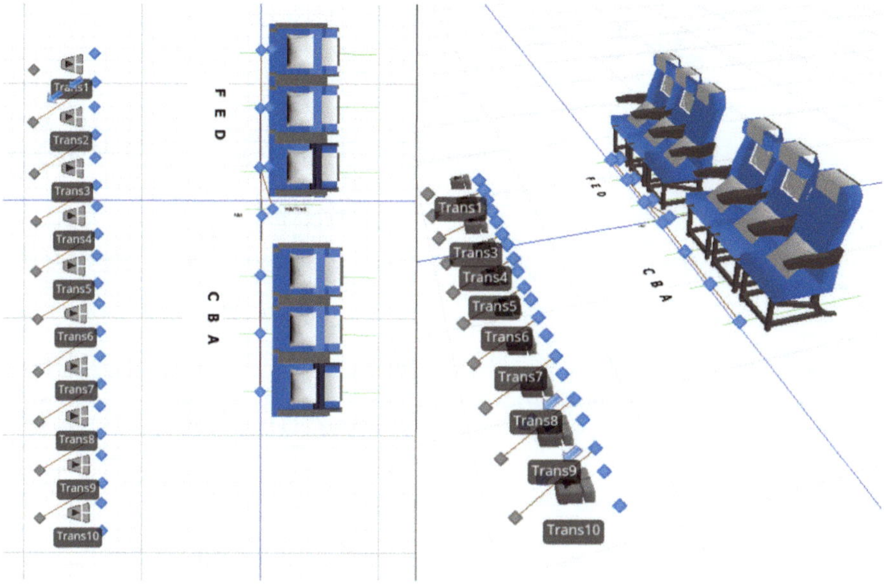

Fig. 4.25 The elements of the DES software model

Fig. 4.26 Station elements

figure the *Trans* objects correspond to transitions 1 to 10 and the logic of those transitions are implemented using the *processes* window of SIMIO.

Other SIMIO elements are used to model the place nodes and transition nodes. For the place nodes, the *Station* elements are the natural element for holding the entities (passengers) and their status are used in the evaluation of the transition semantics of the CPN models. Figure 4.26 presents the different stations used in the object; some of those stations are just used to store the entities that simulate the passengers sit in the cabin seats. Other stations, namely HOLD and PAX, are used to represent the place nodes AISLE and PAX respectively of the CPN model.

Figure 4.27 is an example of the logic for transition SIT6 (Transition6) of Fig. 4.23. First the step DECIDE evaluates if there is a passenger waiting for a seat

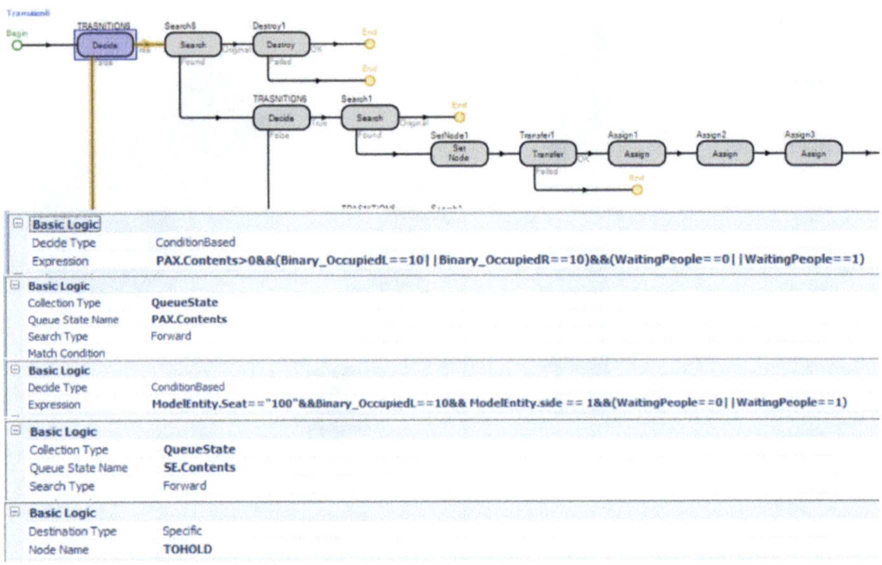

Fig. 4.27 The CPN logic coded in SIMIO steps

(PAX.Contents>0) and if the seat is occupied at the middle (Binary_OccupiedL=010|| Binary_OccupiedR=010) and that there is either no one or only one passenger waiting for a seat (WaitingPeople==0||WaitingPeople==1).If the previous conditions are fulfilled then the next condition checks whether the passenger goes to the left side (ModelEntity.side==1) and that the passenger needs the window (ModelEntity.Seat==100).

The second SEARCH step looks for the contents in the SE station (see Fig. 4.27) accessing the Queue called SE.contents. If there is one element in the queue then it means that the passenger is blocking and then through the set node and the transfer node it is sent out of his seat. The remaining steps are used to update the values of the different variables used.

A similar coding is performed for all the different transitions of the model and they are coded for developing the module that represents the row of the cabin. Once the module of the row is developed, the whole cabin is constructed by making instantiations of the module and connecting them together using the capabilities of SIMIO. The top part of Fig. 4.28 illustrates the whole model of a cabin once the different instantiations of the row module are put together. Every time an entity (passenger) enters to a module, the CPN logic behind the model will steer the simulation while the rest of the time the dynamics will be governed by the SIMIO DES engine thus making a more robust simulation. In Fig. 4.28 the final result of the cabin simulator developed using this approach is presented together with some snapshots of the simulation. In particular the snapshots illustrate the situation when the passengers need to get out so that the arriving passenger reaches his place at the window, and it also shows in the last snapshot that while the

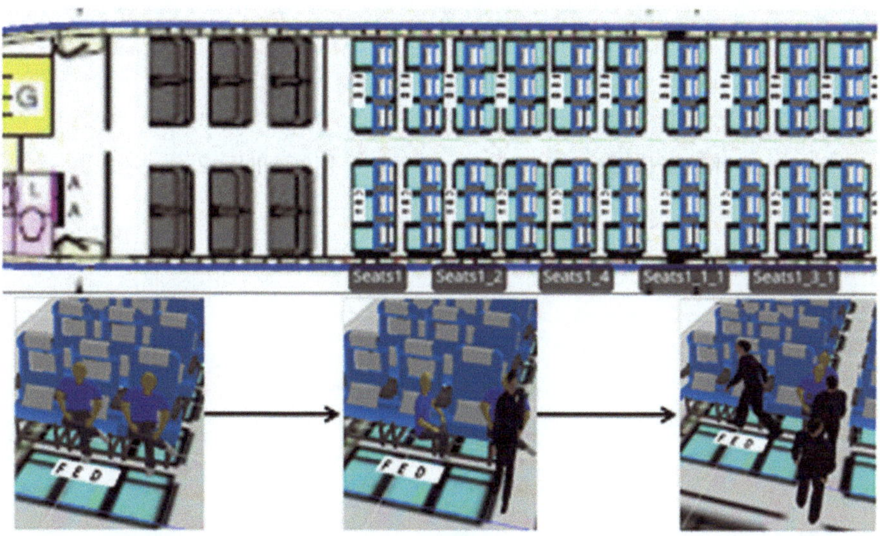

Fig. 4.28 The different modules put together

passengers are letting the arriving one passes to get his seat they are blocking the aisle thus generating a queue as a consequence of these events (emergent dynamics).

4.5.2 Example 2: Sequential Manufacturing System

The following example illustrates the modelling of a production of goods in which every product follows a different sequence using different machines, it is also known as a job-shop. This type of system could be a production line of products, the one in a chemical plant or even the process for making beer which is very popular nowadays. Figure 4.29 illustrates the type of system that I have just mentioned. The circles represent the raw materials and the triangles the produced good. As it can be seen the different goods follow a different sequence within the system and the challenge in these types of systems is to come with an efficient schedule or resources (machines) since there is competition for the resources among the different products.

In this example, the production system is modelled by exploiting the CPNs' capacity for abstraction, so we modelled it using only 2 place nodes and 1 transition node. However, the modeller could develop it in a more disaggregated way as we presented in the previous example.

Figure 4.30 depicts the model developed. In this example the tokens have three colours that are used to specify some characteristics that will be described in the following paragraph. Place node P2 models the availability of machines to carry out

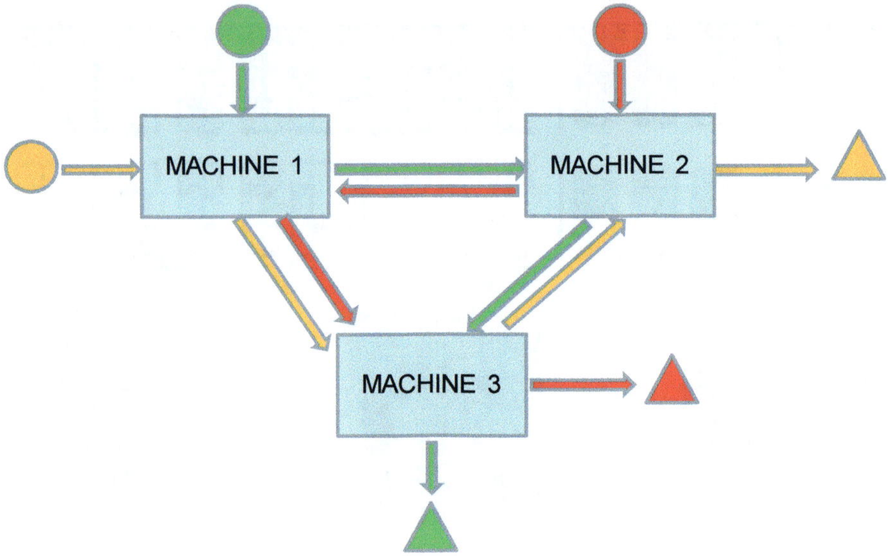

Fig. 4.29 The job shop production system

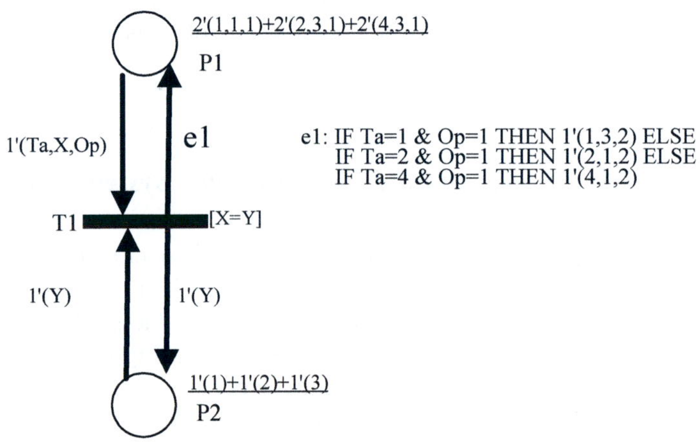

Fig. 4.30 CPN model of a manufacturing system

operations; the colour of the tokens in this place node corresponds to the availability of machines.

The information that governs the evolution of the model is coded in the colours of the model's tokens. Table 4.3 gives a description of the colours used in the model.

The existing place nodes model, on the one hand, the type of activity, machine requirement and sequence of operations of the job to be done (place P1) and, on the

Table 4.3 Description of the colours of the CPN model

Colour	Domain	Description
Ta	Integers	Describe the number of the task to be done
X	[1..3]	Describe the number of the machine required to carry out an operation
Op	[1..2]	Describes the next number of operation for a specific task
Y	[1..3]	Identifier of the available machine

Table 4.4 Description of the colour sets

Place	Colour set	Description
P1	Product (Ta*X*Op)	This group describes the relationship between tasks to be done with their corresponding operation and the machine required to carry out the operation
P2	Y	Describes the availability of machines for the corresponding tasks to be done

other, the availability of processing machines (place P2). Table 4.4 illustrates the colour sets used for the two corresponding places of the CPN model.

When the colours have been defined, it is important to define the arc expressions and the guards associated with the model. The arc expressions are described in Table 4.5.

Transition T1 represents the execution of an operation; we keep track of the sequence of operations by means of the colours of the model's tokens. In this example the *Guard* specifies that the number of available machine matches the one required by the process specified by the second colour of the tokens in P1. In this example the expression *e1* of the output arc in T1 updates the sequence required by the different entities.

4.5.3 Modelling of Place Nodes in SIMIO

As mentioned previously, the modelling of the place nodes is done using the *station* element that will make a correspondence to place nodes P1 and P2. Figure 4.31 illustrates the definition of the two stations that will take the role of place nodes P1 and P2 in the model of Fig. 4.30.

Because these stations have no standalone object for representing them in the animated model, the graphic representation will be done making use of a pair of *detached queue* in order to be able to display the number of entities that are in the place nodes during the simulation run. Figure 4.32 illustrates how the queues are assigned to the corresponding station. We need to add the object *queue* and then specify in its properties window to which station the queue is attached to.

It is important to note that, if we want to display the entities in the station, we have to specify the name of the station in the property that refers to the queue state.

Table 4.5 Description of the arc functions

Type of arc	Expression	Description
Arc (P1,T1)	1'(Ta,X,Op)	Selects a token within the place node from those available and assigns their colours to the variables of the arc expression
Arc (T1,P1)	IF Ta=1&Op=1 THEN 1'(1,3,2) ELSE If Ta=2 & Op=1 THEN 1'(2,1,2) ELSE IF Ta=4&Op=1 THEN 1'(4,1,2)	Establishes the logic of the tokens that will be generated to P1 with the firing taking into account the previously performed operations
Arc (P2,T1)	1'(Y)	Selects a token from place node P2, which represents available machines
Arc (T1,P2)	1'(Y)	Returns the token used previously so that it can be used again

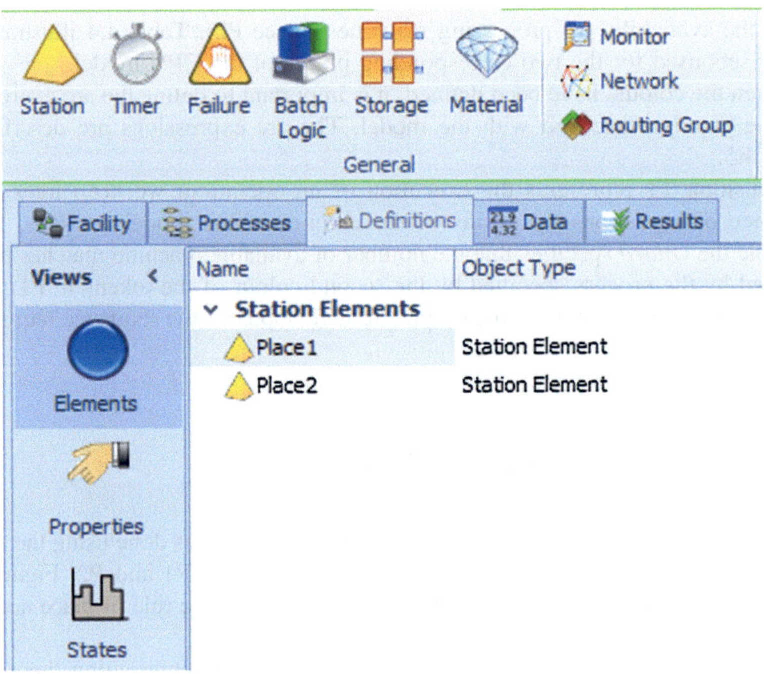

Fig. 4.31 Definition of places

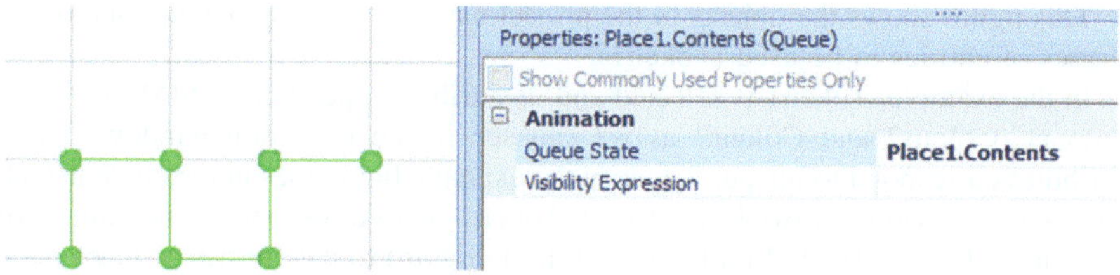

Fig. 4.32 Graphic assignments of queues

Using the '.' Operator we specify the field of the corresponding property that, in this case refers to the contents in the place:

Place1.Contents

With these *queue* objects it is possible to visualize the entities within the stations, however this is not mandatory and the stations would work in the background during the simulation with the difference that the user will not be able to visualize the movement.

4.5.4 Definition of Token Colours

We will use two types of SIMIO entities in this example for modelling the two types of tokens that participate in the CPN model. We specify the attributes associated with these entities when selecting the corresponding entity in the project navigator and accessing the *Definition* window as it is illustrated in Fig. 4.33.

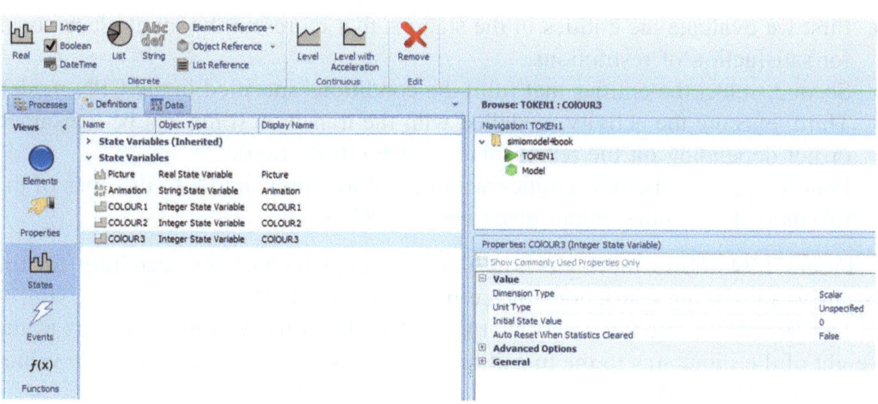

Fig. 4.33 Definition of attributes

This figure shows the objects in the project navigator: the model and one of the classes of entities to be used (Token1).

In the example, Token1 is selected and its attributes specified as SIMIO's states Colour1, Colour2 and Colour3 as we can see in the definition window. These attributes correspond to integer values in the example but it does not exclude that all types can be used (e.g. Boolean, string). Token1 will correspond to the entities of place node P1 and Token2 (not depicted in the figure) to the entities of place node P2, in this example Token2 will have only one colour.

4.5.5 Modelling Transitions

As mentioned in the above section, we use different steps of the process window such as *Decide* and *Search* to assess the conditions to be satisfied by the tokens in the place nodes in order to fire a transition.

The standard conditions to be satisfied are:

- The number of tokens in the entry nodes must be equal to or more than the weight of the input arc.
- The colour of the corresponding tokens must satisfy the corresponding value of the input arc expression.
- The combination of colours must satisfy the Boolean expression of the *Guard* associated with the transition.

We will use other steps such as *assign*, *setnode*, and *destroy* to specify the logic when the entities do not satisfy the constraints or when we need to change the value of the colours because of the arc expression associated with the output arc of the CPN model.

During the implementation we perform four basic steps:

- First we evaluate the entities in the stations that represent the input place nodes for satisfaction of restrictions.
- Second, select the entities that fulfil the conditions specified in the CPN model.
- Third, execute the activity modelled with the transition (time can be consumed or not depending on the representation of the transition).
- Fourth, generate the new entities with the information similar to the ones of the tokens that go to the output place nodes of CPN.

Figure 4.34 shows the logic flow that was implemented for searching for the tokens to satisfy the restrictions of the model in Fig. 4.30:

The *Decide1* is used to verify that the entry places have as many tokens as the weight of the input arcs to the transition. This is coded using the *expression* values specified in Fig. 4.35.

The figure shows the properties window of the *Decide*1 step. In order to carry out the specification of the constraint, the *Decide*1 type will be set to

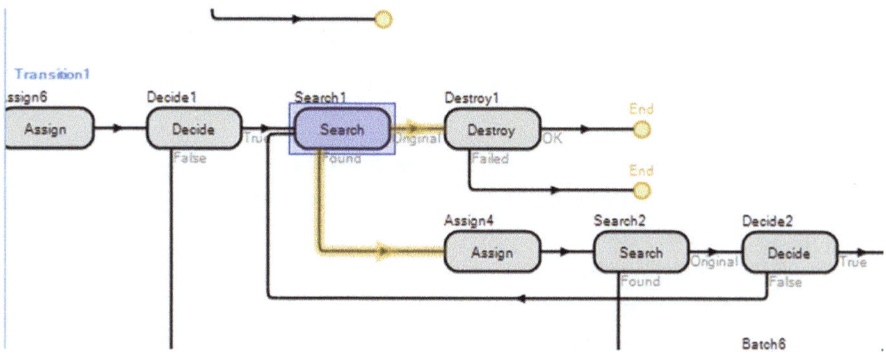

Fig. 4.34 Implementation of the logic of the transitions

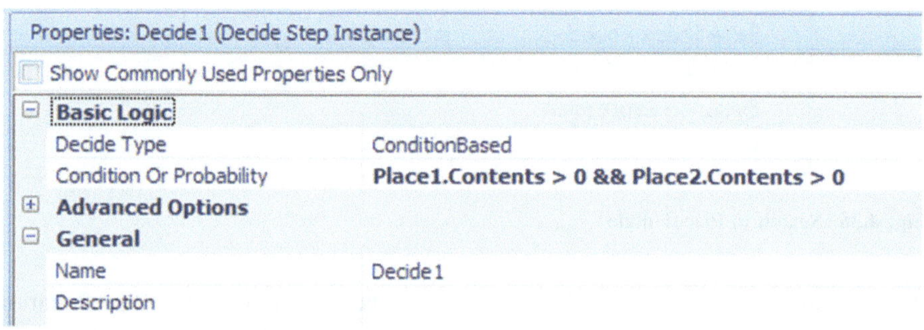

Fig. 4.35 Satisfaction of restrictions

ConditionBased; otherwise the logic will be ruled by probability of occurrence specified in the corresponding field which is not our case. The following expression assesses the existence of sufficient tokens:

$$\mathbf{Place1.Contents > 0\&\&Place2.contents > 0}$$

As it can be observed in Fig. 4.34, if the condition assessed is true, the flow of the process continues on the *true* side of the exit from *Decide1* step.

Assuming that the outcome from *Decide1* is True, the following *Search* step will be used to look in the stations for those entities that satisfy the characteristics specified by the constant values of the arc expressions of the CPN model. This is accomplished by using the *Match Condition* property which can be found in the properties windows of the step *Search*. This property is utilized for making a search, using specific data, in a set of objects (in this case the content of Place1 entities), such as constant values or logic conditions (Fig. 4.36).

In this example, the *Search* chooses any entity from the internal list of the station. In our example of CPN, the input arc coming from P1 requires only the assignment of values to the variables Ta, X, and Op; so there is no real constraint on

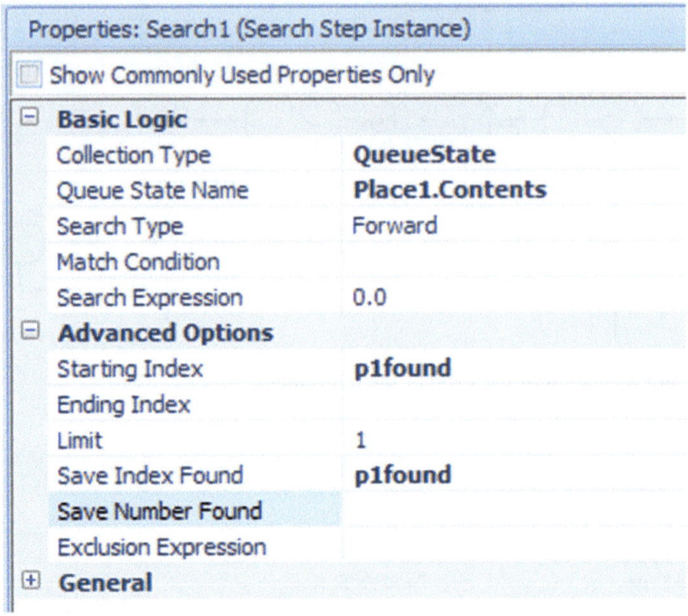

Fig. 4.36 Search in Place1 node

the selection of an entity from this place node. The search inside the Place2 station will therefore be constrained by the values specified by the aforementioned variables. In our case, we will use the property *Save index found* in order to have in computer memory the index of the entity that has been selected by the search.

Assuming that we found an entity that fulfilled the previous criteria, the next step in the assessment of Fig. 4.34 sequence is *Assign4* step, which is used for assigning the variable X, the value of the second colour of the entity selected, as can be illustrated in Fig. 4.37.

When the value of variable X has been assigned, which, due to the guard associated with the CPN model, is the only constraint to be satisfied, a new search is done with step *Search2* on the set of entities belonging to *Place2* station. The

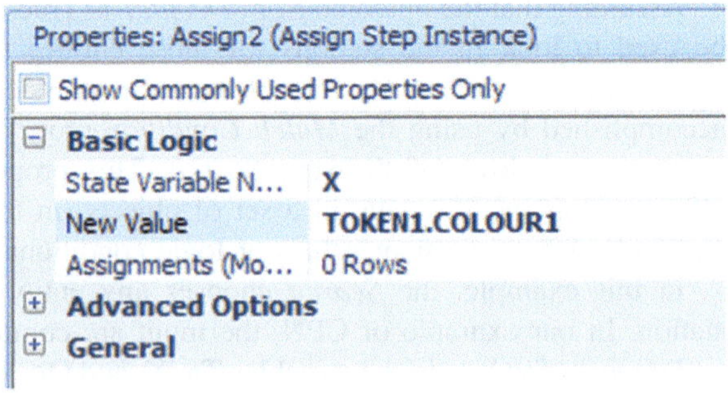

Fig. 4.37 Assignment of values to variables

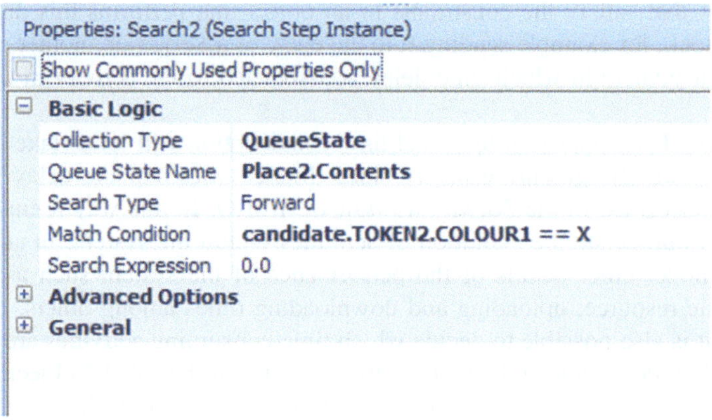

Fig. 4.38 GUARD conditions

preceding operation will now be done by specifying in the *Match Condition* property that it is constrained by the value of variable X previously assigned. *COLOUR1* of the entity from place2 must match the value of X as it is specified in the guard of Fig. 4.30 node (Guard condition). The guard restriction is implemented in the next *Search* as it is illustrated in Fig. 4.38.

The properties include *Return Value*, which will be used to store the search result in the computer memory (1 means that it found an entity with the characteristics that the search required, while 0 means that it did not). If the search finds an entity that satisfies the constraints (weight of the arc, known value of arc and guard expression), the logic flow continues through the *Found* node of the *Search* step, which associates the *Found* with the entity found in the search. The entities that satisfy the constraint must now be sent into the SIMIO model. For this objective we use either a *Setnode* or *Transfer* step for sending them to the required location.

On the other hand, if the search has not produced any entity that complies with the Boolean condition X = Y, another entity would have to be chosen from the *Place1* node and the process is repeated all over again.

Figure 4.34 shows that the *False* exit of the *Decide2* step directs to the initial *Search* step. This means that the search is performed from the index of the last entity found. That is why the index of the first entity found is stored using the *p1 found index* (Fig. 4.36). This way we perform a thorough search through the set of entities that satisfy the constraints.

4.5.6 Time Consumption

There are two ways of modelling the time consumed while executing the activity. The first way is the one mentioned in the previous section and consists of adding a *Delay* step which causes a delay in the model's clock. The second option is to send the entities that satisfy the constraints to an object that performs this delay in the model's clock, for example sending it to the entry of a *Server* or another object like a *Path* or a *Vehicle* in which time delay is associated to the activity in the SIMIO model.

Using the latter approach to model time consumption will let us take advantage of SIMIO's 2D-3D graphic capacity since the activities and the delay related to them are performed in the *Facility window* of SIMIO. In this way it enables us to graphically appreciate the evolution of activities within the system. In addition we can also model other details of the performance of the system such as: possibly faults in the resource; uploading and downloading times among others. Using this approach it is also possible to decide which time-consuming activities are modelled by the CPN model logic and which by the activities of the SIMIO object; therefore they can be disaggregated by the modeller to have better control.

In the example being developed, the found entity will be sent to a *Server* object. In the same example, the corresponding process time (*Delay*) can be modelled in a deterministic or stochastic fashion, or depending on some property of the entity. In this case, the time consumed will depend on the **Ta** value of the attribute of the *Place1* node token, as seen in Fig. 4.39.

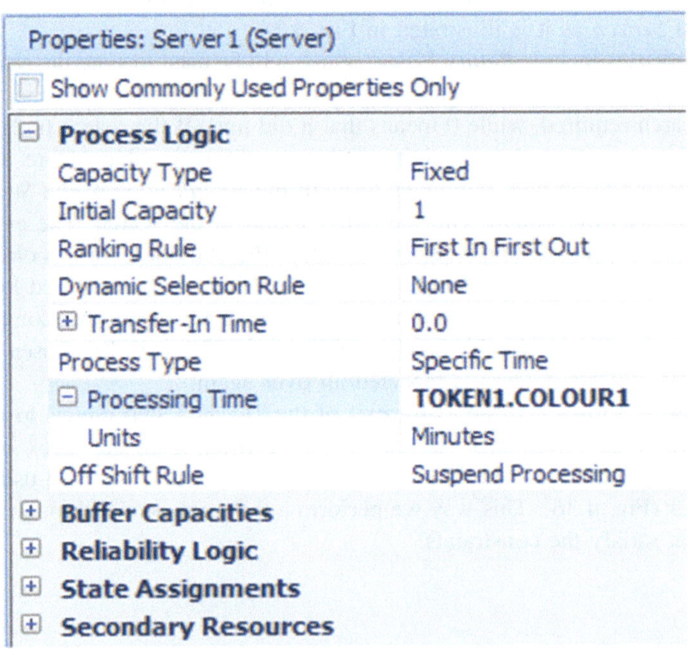

Fig. 4.39 Assigning time consumption to the server object

4.5.7 Modelling Exit Arcs

The remaining element for the implementation of the CPN in this environment is the output arc or arcs. The coding of the functions of the output arcs is even simpler than the corresponding coding for the transitions, which only requires the use of *Assign* and *Decide* steps. The values of the entities exiting the transition are assessed with the *Decide* step and this is used as the basis for assigning values to the colours of the tokens, in accordance with the logic of the CPN model. Figure 4.40 shows the implementation developed for the example being analysed in this section.

In this example, the function of the exit arc corresponding to ***e1*** (of the CPN model) is called *OutputFunction* and will be coded through a combination of nested *Decide* and *Assign* steps. The first *Decide* verifies if the following expression holds:

$$\textbf{Token1.COLOR1} == \textbf{1\&\&Token1.COLOR3} == \textbf{1}$$

If the result of the assessment is positive, the flow continues through the *True* exit of the *Decide1* step and the following *Assign*s are executed in order to assign the new values to the tokens' attributes with the following values:

Assign1	**Token1.COLOR2==3**
Assign2	**Token1.COLOR3==2**

The rest of the *Decide and Assign* steps perform the same operation, but just for the other possible results of the assessment of function ***e1***.

It should be mentioned that these arc functions must be inserted in the SIMIO model just after the activity which performs the correspondent delay (a process, a server, etc.) is executed. In this example if, having satisfied the associated

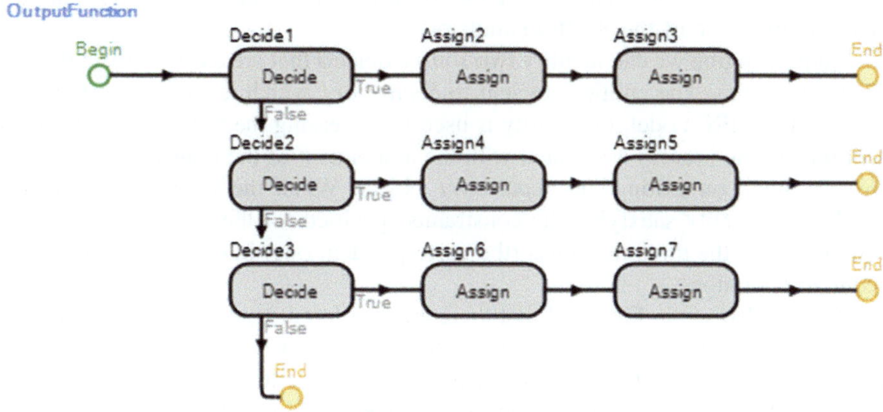

Fig. 4.40 Coding of exit arc functionalities in the process window

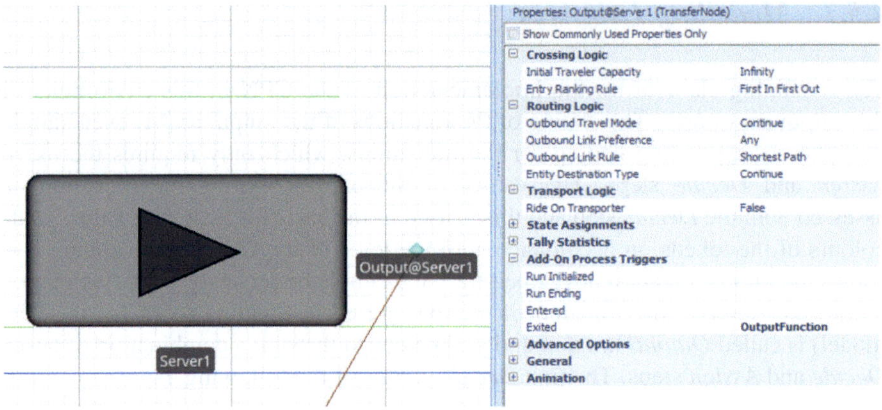

Fig. 4.41 Association of exit arcs to the add-on triggers

constraints, the entities are sent to a server to model the time consumption, the exit arcs must be associated with a server event just after the activity that suffers the delay ends, or right after the delay in the simulation clock. In the example described here, the association of the exit arcs is done in the server as soon as the entity exits from the server (using Add-On Triggers), Fig. 4.41 illustrates the implementation.

The association is implemented on the transfer node at the *Server* exit. Therefore in this example the delay is modelled by the use of a server of SIMIO.

4.5.8 Final Model

Figure 4.42 shows the final version of the implemented example illustrating the points where the entities go *out* of the SIMIO model and enter to the domain of CPN semantics and we can also see the locations where the entities go *In* and are ruled by the logic of the SIMIO simulator.

From the figure we can identify two *source* objects that are used to create the two types of entities. The entities are sent to *Separator1* which is used to perform the logic of the CPN model. One entity is used for assessing the transition. This entity will initiate the process associated with the assessment of the transition on the first *transfer node* on exiting the *Separator1* object. When the transition evaluation results in an entity satisfying the constraints specified for the model, this entity is sent to one of the three machines (the correspondent one) in order to carry out the required operation.

We can also see in Fig. 4.42 that there are 3 different machines for carrying out three different operations. Making use of the *Setnode* and *Transfer* steps, the entities are sent to the node identified as a *transition* in order to then be directed to the corresponding machines. On exiting the corresponding operations, the logic of the exit arcs must be associated with some event after visiting the machines as it has

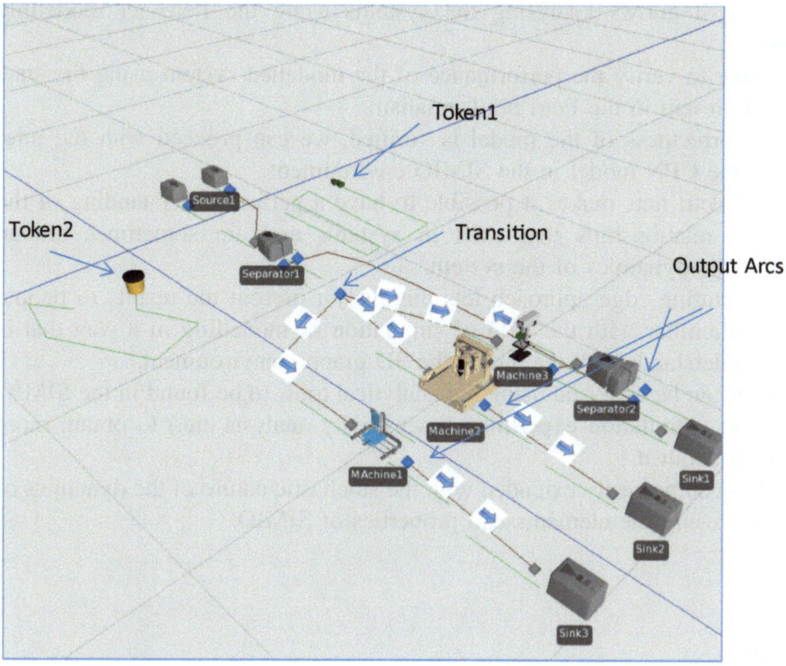

Fig. 4.42 Final CPN model with the SIMIO environment

been previously mentioned. The example shows that this functionality will be associated with an event of the process *exit node* in the corresponding machine.

Lastly, we use a *Separator* object just as an auxiliary object that sends a copy of the entities to the station that models the *place1* node and another copy will be used to send the entity to the *Separator1* object for re-assessing Transition1 when the operation of the corresponding machine is finished. This way we can see that the original system will be suitably modelled and the flow of entities is governed in accordance with the semantics of CPNs making use of the power of SIMIO at the same time. The above implementation allows taking advantage of the modeller's capacity while exploiting the full potential of SIMIO in a unambiguous fashion thus making a more robust model that is useful for getting an adequate analysis of the dynamics of the original system.

4.6 Conclusion

This chapter presented how to integrate the modelling formalism known as coloured Petri nets in SIMIO, a commercial DES-based simulation environment. There are many advantages of integrating the environment with formalism:

- First of all, it allows analysing the systems using the Petri-net modelling formalism.
- It is possible to verify the performance of the modelled system using the state space tool present in the Petri net formalism.
- Once the correctness of the model is verified, we can proceed with the integration of the CPN model in the SIMIO environment.
- The use of Petri nets makes it possible to have a perfect understanding of the cause-effect relationships that exist in systems and are sometimes hidden because of the dynamics of the system.
- Using this cutting-edge approach lets the analyst present the results to people who are unfamiliar with the field of simulation or modelling in a way that is easier to understand, with the aid of the 3D graphic environment.
- The system can be analysed using the analytical tools to be found in the SIMIO environment (simulation experiments, sensitivity analysis etc.) to obtain more information about it.
- The models can be easily extended with the stochastic nature of the dynamics of the systems using the elements and properties of SIMIO.

References

Aarhus University. (2016). WebPage. http://www.au.dk/en/

CPN Tools WebPage. (2016). http://cpntools.org

Jensen, K. (1997). *Coloured petri nets: Basic concepts, analysis methods and practical use*. Berlin: Springer.

Jensen, K., Kristensen, L. M. (2009). *Coloured petri nets: Modelling and validation of concurrent systems*. Berlin: Springer.

Kelton, W. D., Smith, J. S., Sturrock, D. T., & Verbraeck, A. (2010). *Simio and simulation: Modeling, analysis, applications*. Boston: McGraw-Hill.

Mujica, M., Flores, I. (2015). CPN-DES modular simulator for assessing boarding performance of aircraft. *Wintersim 2015*, Case Study Presentation, Huntington Beach, CA, USA.

Mujica, M., Piera, M. A. (2011). Integrating timed coloured Petri net models in the SIMIO simulation environment. In *Proceedings of the 2011 Summer Computer Simulation Conference* (pp. 91–98).

Pegden, C. D. (2007). SIMIO: A new simulation system based on intelligent objects. In *Proceeding of the 39th Winter Simulation Conference*. Sewickley, PA: Simio Corporation.

Weisfeld, M. (2009). *The object-oriented thought process* (3rd ed.). USA: Pearson Education.

Westergaard, M. (2016). Webpage. https://westergaard.eu/

Chapter 5
Simulation Examples

Antoni Guasch and Jaume Figueras

5.1 Introduction

Chapter 1 introduced the phases of a simulation project, emphasizing the importance of each one of them. In consequence, it would be a serious mistake to think that a simulation project is simply coding the model by using simulation software and then extracting the results.

Chapter 2 reviewed the important statistical aspects of simulation, and then Chap. 3 introduced Petri nets as a method for conceptual modeling of the process involved. Many simulation reference books put particular emphasis on the statistical aspects, for both the parameterization of the model and the validation, experimentation and analysis of the results. This book highlights the importance of conceptual modeling as a step prior to the construction of a good Simulation Model. This chapter gives a set of examples initially modeled using Petri nets and later coded with Simio.

One of the most complex phases of a simulation project is the validation of the model, in other words, proving that the model behaves like the real process involved. The best way to approximate the validation step is to suppose that the model is incorrect and do everything necessary to try to demonstrate this. If we cannot demonstrate that it is wrong, we can assume that the model is valid, although the possibility always exists that it is not. One of the mechanisms that can be used to try to validate the model is to contrast the results of a model with the results of another model constructed using different techniques and, if possible, using different work equipment. If the results coincide, the conviction that the simulation model is valid is reinforced in the conclusions.

Whatever the validation method used, we recommend doing a theoretical analysis of the process prior to simulation. In the simple examples given in this chapter, the results of the theoretical calculations would be expected to be very similar to those obtained by simulation. The differences between the results of the two models should be justified based on the different hypotheses used in the

I.F. De La Mota et al., *Robust Modelling and Simulation*,
DOI 10.1007/978-3-319-53321-6_5

construction of each model. If there is no justification, the differences could be a sign of an invalid simulation model or of wrong theoretical calculations.

5.2 Canal-Lock System

Ten barges are used to transport materials in a canal. The barges are loaded by one of the two cranes located on the low part of the canal and are unloaded by one of the two cranes to be found at the end of the high part of the canal. Therefore, two barges can be loaded or unloaded simultaneously, one at each end of the canal. The canal is divided into two sections, the low section and the high one, which are separated by a lock that can only raise or lower one barge at a time. These ten barges only work on the canal and never leave it, using the canal as if it were a closed circuit. At the start, it is assumed that the 10 barges are in the low part, queued for loading. Figure 5.1 shows how the lock works.

Table 5.1 shows the triangular distribution of the times associated with the different operations.

It is a good idea to do a theoretical analysis of the process before the simulation study.

In order to do the theoretical analysis, we will employ basic queue theory concepts. The parameters of a queuing process are:

- λ mean frequency of arrivals: With a known (expected) mean time between arrivals $E(A)$, the mean frequency of arrivals is calculated as $\lambda = 1/E(A)$.
- ω mean frequency of service: With a known (expected) mean time of service $E(S)$, the mean frequency of service is calculated as $\omega = 1/E(S)$.
- s number of servers.

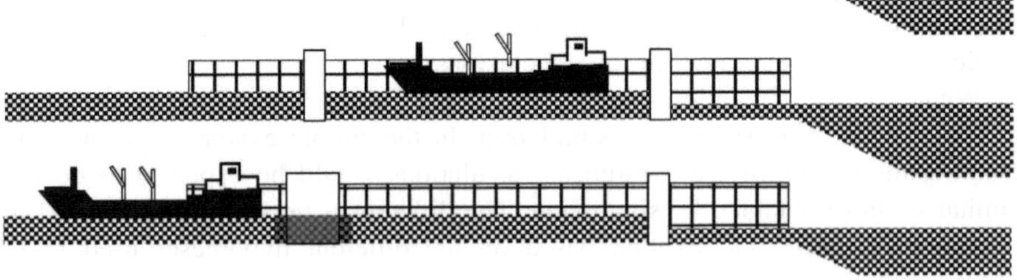

Fig. 5.1 Operational schema of how a lock works

Table 5.1 Process time (triangularly distributed and expressed in hours)

			Minimum	Mode	Maximum
Loading time	2.00			2.25	3.00
Unloading time	1.50			2.25	2.50
Lock time	0.40			0.50	0.6
Path of the low section					
	Raised	2.00		2.50	3.00
	Lowered	1.50		1.75	2.00
Path of the high section					
	Raised	1.00		1.25	1.50
	Lowered	0.50		0.75	1.00

The server utilization factor can be defined based on the aforementioned parameters:

$$\text{Server utilization factor}: \ \rho = \frac{\lambda}{\omega \cdot s} = \frac{E(S)}{E(A) \cdot s}$$

In this example we want to calculate the average utilization factor of the loading and unloading cranes as well as of the lock. For this calculation, we use the mean operating times of the above triangular distributions. The mean is calculated as (minimum + mode + maximum)/3. We do not know the mean time expected between the arrivals of the barges *E*(*A*), but given that the 10 barges work in a closed circuit, we can hypothesize that *E*(*A*) is equal to the time of one complete cycle of the barge divided by the number of barges

$$E(A) = \frac{11.75}{10} = 1.175$$

This makes it possible to calculate the utilization of the loading cranes:

$$\rho = \frac{\lambda}{w \cdot s} = \frac{E(S)}{E(A) \cdot s} = \frac{2.416}{1.175 \cdot 2} = 1.02$$

The value of 1.02 indicates that with this hypothesis, the loading cranes will work at 100% and, therefore, they will limit the barges' arrival rate to the remaining resources. Thus, it seems more correct to suppose that the arrivals rate is the mean time of service *E*(*s*) divided by the number of loading cranes

$$E(A) = \frac{E(s)}{s} = \frac{2.416}{2} = 1.208$$

And the utilization of the resources is:

$$\text{Loading cranes}: \rho = \frac{\lambda}{w \cdot s} = \frac{E(S)}{E(A) \cdot s} = \frac{2.416}{1.208 \cdot 2} = 1$$

$$\text{Unloading cranes}: \rho = \frac{\lambda}{w \cdot s} = \frac{E(S)}{E(A) \cdot s} = \frac{2.083}{1.208 \cdot 2} = 0.86$$

$$\text{Lock}: \rho = \frac{\lambda}{w \cdot s} = \frac{E(S)}{E(A) \cdot s} = \frac{0.5 \cdot 2}{1.208 \cdot 1} = 0.83$$

Note that the lock works twice for every cycle of the barge.

Once the theoretical analysis has been done, we can continue with the simulation study. Before coding in Simio, we think it advisable to obtain the conceptual model of the process in Petri nets, given in Fig. 5.2.

If we look at the Petri net shown below, we note that the barges are not treated like resources. The barges are the temporary entities that flow through the process. The developed model is aimed at the process where the rising passage and falling passage through the lock are differentiated.

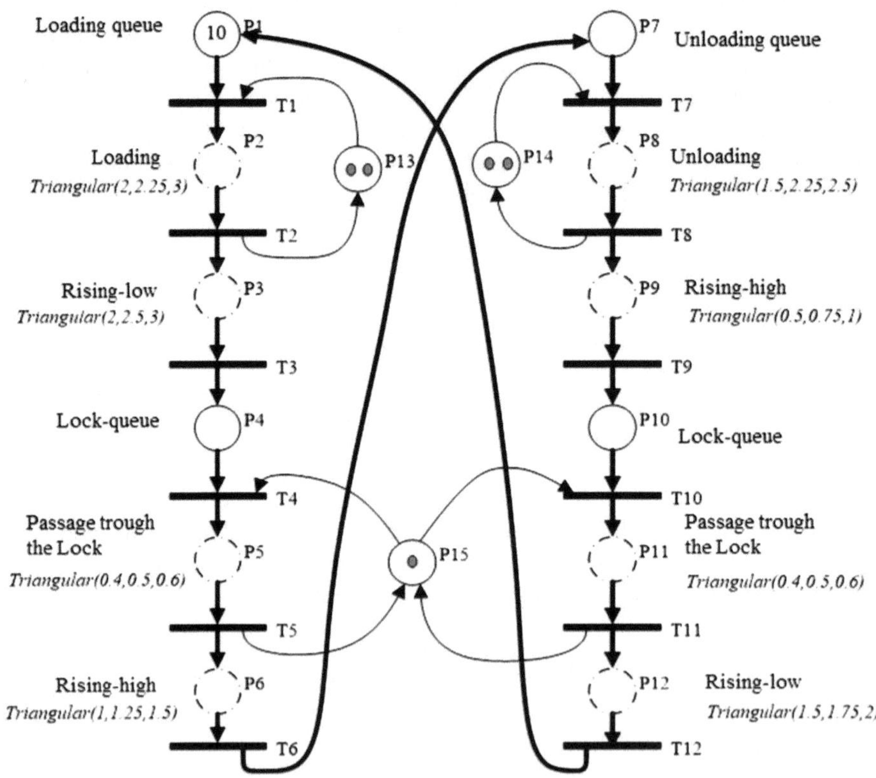

Fig. 5.2 Petri net model for the canal-lock system

With the places being:

- P1: queue of barges awaiting loading
- P2: loading activity
- P3: rising activity of the low section
- P4: queue in order to access the lock in the lowering direction
- P5: activity of passage through the lock in the rising direction
- P6: rising activity of the high section
- P7: queue of barges awaiting unloading
- P8: unloading activity
- P9: lowering activity of the high section
- P10: queue in order to access the lock in the lowering direction
- P11: activity of passage through the lock in the lowering direction
- P12: lowering activity of the low section

And the transitions:

- T1: start of the loading activity
- T2: end of the loading activity and start of the rising of the low section
- T3: end of the rising of the low section in order to queue up to enter the lock
- T4: start of the activity of passage through the lock in the rising direction
- T5: end of the activity of passage through the lock and start of the rising of the high section
- T6: end of the rising of the high section and start of the waiting for unloading
- T7: start of the unloading activity
- T8: end of the unloading activity and start of the rising of the low section
- T9: end of the rising of the low section in order to queue up to enter the lock
- T10: start of the activity of passage through the lock in the rising direction
- T11: end of the activity of passage through the lock and start of the rising of the high section
- T12: end of the rising of the high section and start of the waiting for unloading

Some interesting aspects that have, in a way, already been explained in the previous chapter but that are worth remembering, are:

- The thick lines represent the explicit flow of the temporary entities in Simio.
- Places with an interrupted line represent states whose length of time depends on explicit time functions.
- Places with a continuous line represent waiting states whose duration does not depend directly on the time functions.

Figure 5.3 shows the associated Simio code. The equivalent elements in the Petri net are given for each one of the main objects. The remarks are:

- The object Initial Barges of Source class creates only 10 temporary entities in the initial instant. This operation corresponds to the initialization of the P1 place.
- The movement times in the different sections are coded with objects of the TimePath class.

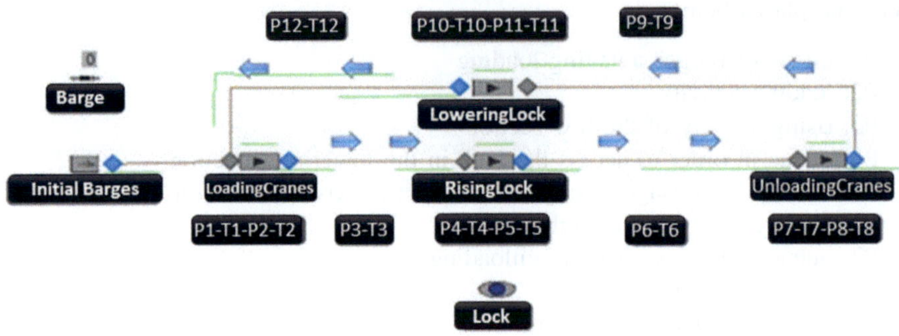

Fig. 5.3 Simio code for the canal-lock system

- The LoadingCranes, UnloadingCranes, LoweringLock and RisingLock objects belong to the Server class. The LoweringLock and RisingLock objects compete for the same Lock resource. This is clearly observed in the Petri net, but is implicit in the coding in Simio.

Ten replications were done with a warm-up time of 20 h for the purpose of having statistics in a permanent regime that make it possible to compare them with the expected theoretical results. The first result that surprises us is the observation that the average utilization of the loading cranes is 0.956, instead of the expected value of 1, because these cranes are a bottleneck in the process. The deviation is because the real cycle time is 11.75 h plus 0.89 h waiting in the queue. If we take into account that the real cycle time is 12.64, the utilization of the loading cranes would have to be:

$$\rho = \frac{\lambda}{w \cdot s} = \frac{E(S)}{E(A) \cdot s} = \frac{2.416}{1.264 \cdot 2} = 0.956$$

I.e., the real cycle time is the factor that determines E(A). Table 5.2 shows the theoretical and real utilization for each one of the resources:

There is no point in increasing the number of barges, given that the system is almost saturated. The utilization factor of the loading cranes is at 95%. If we work with one more barge in the model, we see that the number of transports only rises from 70.6 to 73.6 transports in the complete period of 100 h.

Table 5.2 Utilization of resources

Resource	Theoretical utilization	Real utilization
Loading cranes	1	0.95
Unloading cranes	0.86	0.82
Lock	0.83	0.79

5.3 Two-Robot and 5-Machine Process

Two different types of pieces arrive at the subprocess of a manufacturing plant. These pieces, identified as type P1 and type P2 pieces, follow two different machining processes. See Fig. 5.4. The P1 pieces are processed by being machined by an M1 machine and afterward by an M2 machine; the P2 pieces are only processed in an M2 machine. These pieces arrive at an 8-setting entry buffer and are loaded into the corresponding machine by two robots. The R1 robot loads the M1 machines and the R2 robot loads the M2 machines. The piece (P1 or P2 type) that has been the longest waiting in the buffer is always processed. So, if a P1 type piece arrives, the subprocess follows the sequence given below:

1. If there is space in the buffer, it enters it.
2. The R1 robot takes the piece from the buffer and loads it into an M1 machine.
3. The piece is machined in the M1 machine.
4. The R1 robot leaves the piece in the buffer.

For the P2 type pieces or the P1 type ones already processed in M1, the processing sequence is simpler:

1. The R2 robot takes the P2 type piece or processed P1 type and loads it into an M2 machine.
2. The piece is machined in the M2 machine.
3. The R2 robot leaves the piece at the exit of the subsystem.

The work times are:

- The time between arrivals of P1 type pieces is exponential, with a mean of 10 min
- The time between arrivals of P2 type pieces is exponential, with a mean of 15 min

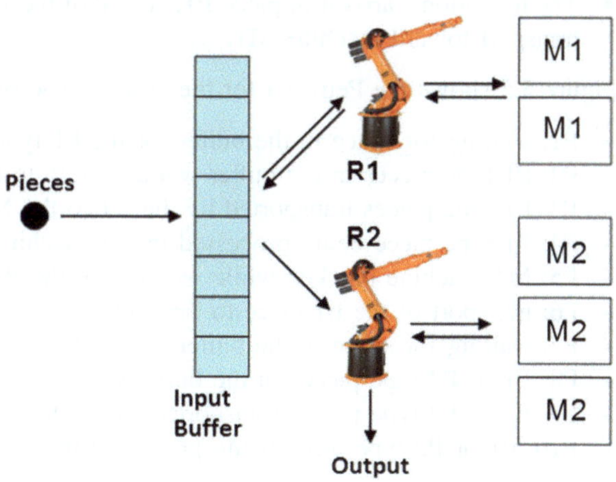

Fig. 5.4 2-robot and 5-machine system

- The total transport time of the robots is 10 s
- The process time of the M1 machine is 18 min
- The process time of the M2 machine is 16 min

And we have two M1 machines and three M2 machines.
The utilization of the resources is

$$\text{Robot1}: \rho = \frac{\lambda}{w \cdot s} = \frac{E(S)}{E(A) \cdot s} = \frac{(10/60) \cdot 2}{10 \cdot 1} = 0.034$$

$$\text{Machines1}: \rho = \frac{\lambda}{w \cdot s} = \frac{E(S)}{E(A) \cdot s} = \frac{18}{10 \cdot 2} = 0.9$$

$$\text{Robot2}: \rho = \frac{\lambda}{w \cdot s} = \frac{E(S)}{E(A) \cdot s} = \frac{(10/60) \cdot 2}{6 \cdot 1} = 0.056$$

$$\text{Machines2}: \rho = \frac{\lambda}{w \cdot s} = \frac{E(S)}{E(A) \cdot s} = \frac{16}{6 \cdot 3} = 0.89$$

The expected time between arrivals at the R2 robot and the M2 machine is

$$E(A) = \frac{10 \cdot 15}{10 + 15} = 6$$

Prior to the Petri net deduction process, it is a good idea to identify:

- The resources (or permanent entities) that are involved in the process: robot R1 (1), robot R2 (1), machine M1 (2), machine M2 (3), spaces in the buffer (8).
- The temporary entities: piece P1, piece P2.
- The activities: transport to M1 machine, process in M1 machine, return to buffer from M1 machine, transport to M2 machine, process in M2 machine and transport from exit of M2 machine in order to leave the subsystem.
- The transitions: arrival of piece P1, arrival of piece P2, entry into buffer, start of transport toward machine M1, …

Figure 5.5 shows the Petri net for the system. The places are:

- P1: waiting for space in the buffer for the P1 type pieces
- P2: P1 type pieces in the buffer waiting to go to the M1
- P3: P1 type pieces transported by the R1 to the M1
- P4: P1 type piece being processed in M1 machine
- P5: M1 machine blocked while waiting for the R1.
- P6: transport of the P1 piece to the buffer
- P7: waiting for space in the buffer for the P2 type pieces
- P8: P1 or P2 type pieces in the buffer waiting to go to the M2
- P9: P1 or P2 type pieces transported by the R2 to the M2
- P10: P1 or P2 type piece being processed in M2 machine

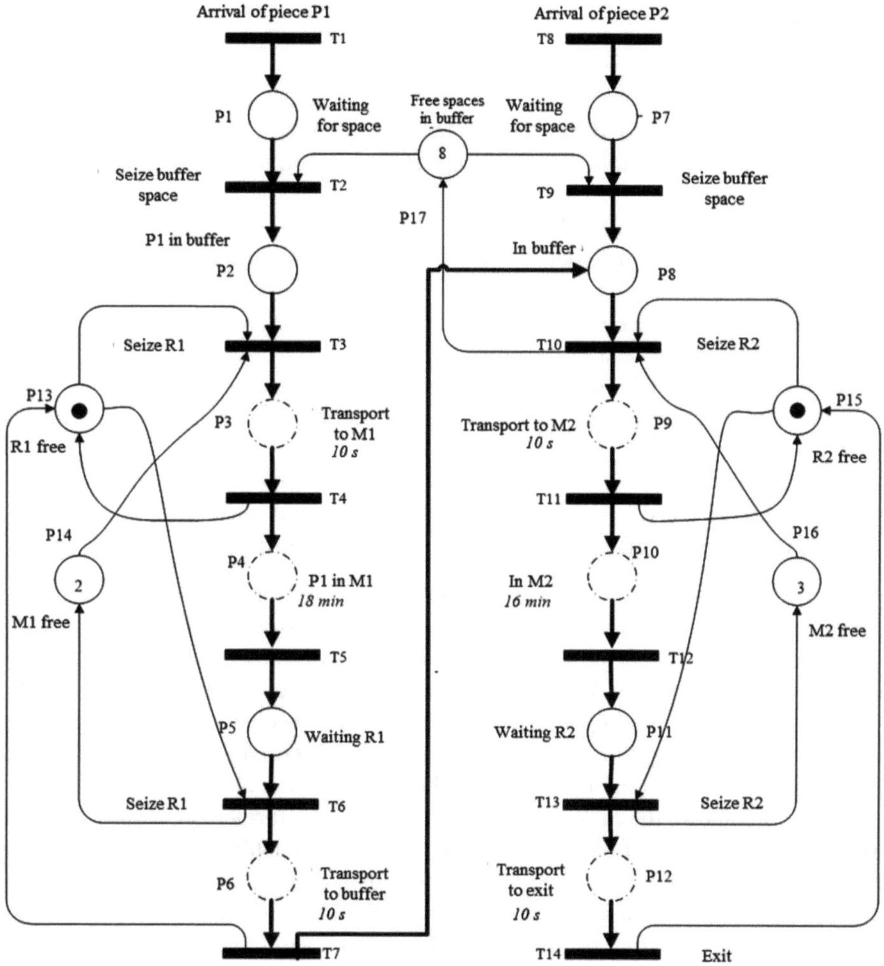

Fig. 5.5 Petri net model for the 2-robot and 5-machine system

- P11: M2 machine blocked while waiting for the R2.
- P12: transport of the P1 or P2 piece to the exit
- P13: R1 free
- P14: M1 free
- P15: R2 free
- P16: M2 free
- P17: Free spaces in the buffer

And the transitions:

- T1: arrival of P1 pieces
- T2: seize of buffer space
- T3: seize of the R1 and M1, before transport begins
- T4: end of the transport, release of the R1 robot and start of the process in the M1
- T5: end of process in the M1
- T6: seize of the E1 in order to return to the buffer and release of the M1
- T7: end of transport with the R1
- T8: arrival of P2 piece
- T9: seize of buffer space
- T10: seize of the R2 and M2 before transport begins
- T11: end of the transport, release of the R2 robot and start of the process in the M2
- T12: end of process in the M2
- T13: seize of the R2 in order to transport to exit and release of the M2
- T14: end of transport with the R2 and exit of the piece

The most important actions in a simulation model are related to the seizing (seize), utilization (delay) and use of resources (release or freeing in a context of scarce resources. The Petri net has the advantage of showing this set of actions explicitly and graphically. In relation to the seizing of resources, we stress that the space is not freed in the buffer when R1 takes the P1 piece; thus, we can guarantee space in the buffer for returning the piece once the process in the M1 is completed. Early release in transition T3 can cause the model notice blocked.

Figure 5.6 shows the Simio code associated with the former net. Transition T3 is triggered if we have a free R1 robot and a free M1 machine synchronously. The coding in Simio of the transitions where more than one resource is seized is a very delicate point, given that in its current version Simio does not in general guarantee the synchronous seizing of resources. This can cause undesirable effects and blockages that are not attributable to the model.

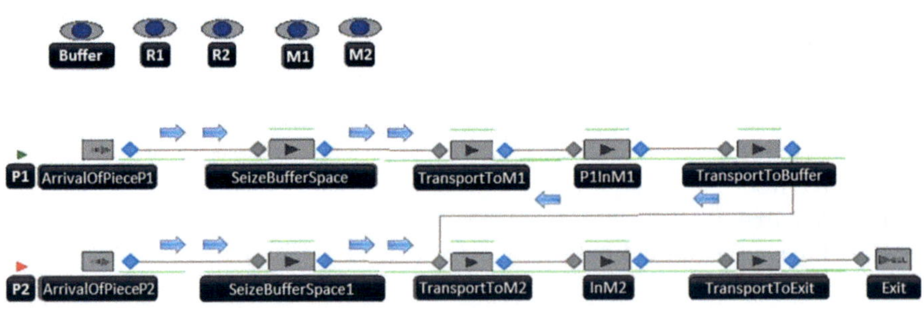

Fig. 5.6 Simio code for the 2-robot and 5-machine system

Figure 5.7 shows that in order to execute the transport it is necessary to seize the M1 machine “On entering” and in “Resources for Processing” robot R1. This coding guarantees that machine M1 is seized first, and robot R1 is seized second. If instead of seizing machine M1 “On entering”, it is seized “Before Processing”, the simulator may, and in fact does, become blocked if it first seizes robot R1 while the two machines M1 are working on other pieces. In order to release one of the two M1 machines, robot R1 has to be free, but robot R1 is seized by the piece that has to be transported to M1.

To conclude, this example stand out because the results of the simulation are very similar to the theoretical results we anticipated for the utilization of the resources.

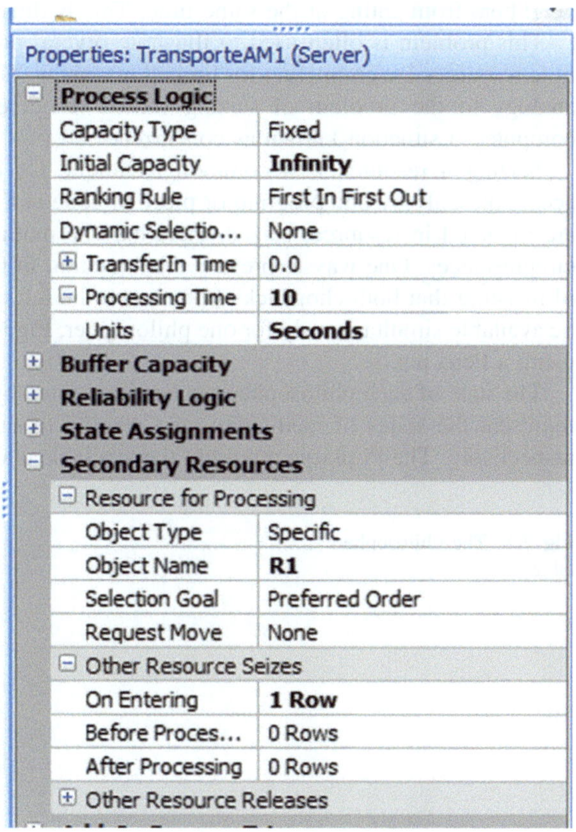

Fig. 5.7 Parameterization of the transporteam1 object of the server class

5.4 The Philosophers' Dinner

The problem of the philosophers' dinner is an illustration of a common problem in concurrent computing and a classic problem of multiprocess synchronization. Dijkstra (1971) established an exam question in a synchronization problem where five computers compete for access to five shared peripheral controller tapes. Very soon after this, Tony Hoare renamed this problem the problem of the philosophers' dinner.[1]

Five Chinese philosophers (f1, f2,…, f5) are seated at a round table. In the center of the table there is a rice bowl. Between each pair of philosophers there is one chopstick. Each philosopher alternates between meditating and eating. In order to eat, the philosopher needs two chopsticks, and he is only allowed to use the two close to him (to his left and right). Sharing the chopsticks in this way stops two neighbors from eating at the same time. This is illustrated graphically in Fig. 5.8.

This problem is often used to illustrate several problems that occur when multiple resources are competing for limited resources. The lack of available forks is an analogy to the problem of seizing shared resources in real programming in a computer, a situation known as concurrence.

Seizing a resource is a common technique for ensuring that the resource is accessible only by one program or piece of code at a time. When several programs are involved in resources seized, there may be mutual blocking depending on the circumstances. One way to prevent this blocking in the case of the dinner consists of insisting that both chopsticks (the one on the left and the one on the right) must be available simultaneously for one philosopher. Figure 5.9 shows a possible model using a Petri net.

The state of each philosopher can be represented by three places (Mi, Ei, Ci) that represent the states of meditation, and waiting to seize the chopsticks and eating respectively. The Pi places represent the available chopsticks. In order to be able to

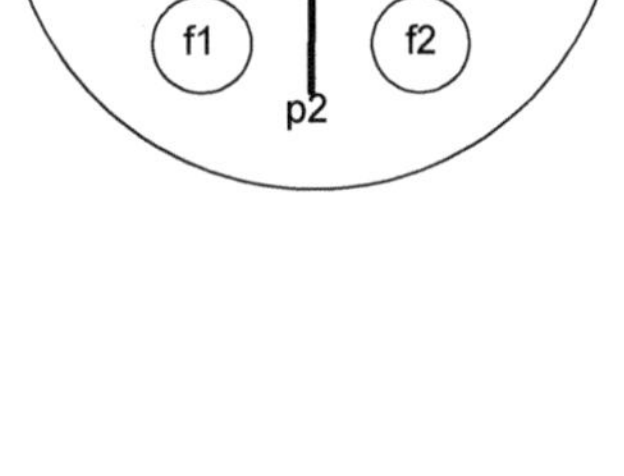

Fig. 5.8 The philosophers' table

[1]See http://en.wikipedia.org/wiki/Dining_philosophers_problem.

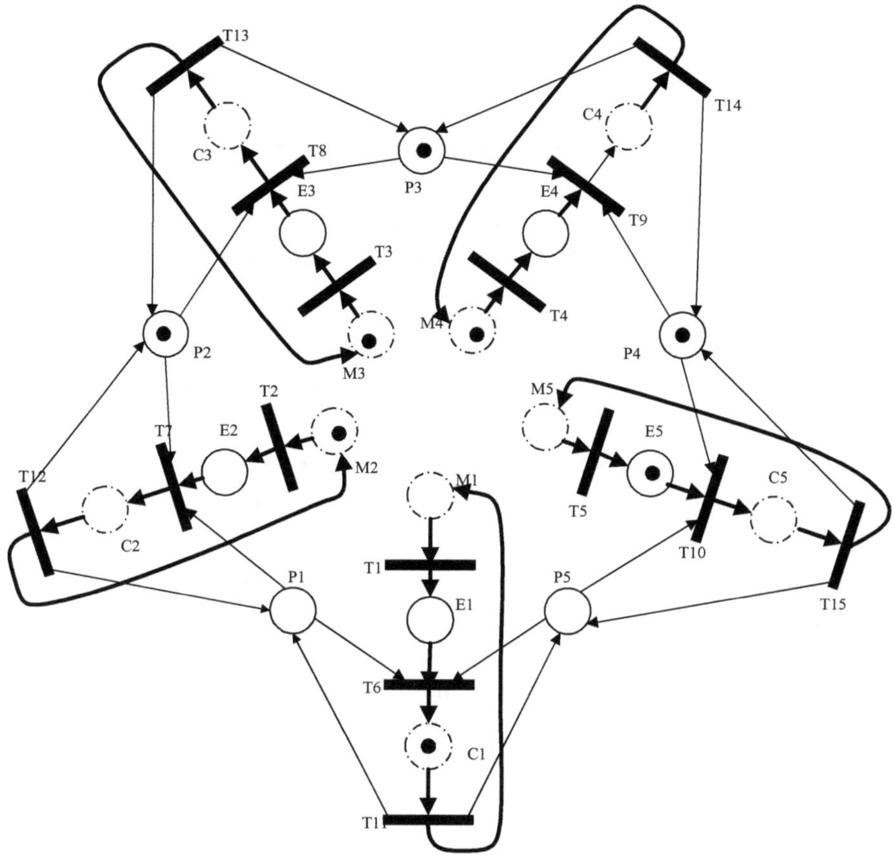

Fig. 5.9 Petri net for the philosophers' dinner problem

pass from the waiting state to that of eating, both chopsticks (the one on the left and the one on the right) have to be available.

The places are:

- M1–M5: meditation of each philosopher
- E1–E5: waiting of each philosopher until left and right chopsticks become available
- C1–C5: philosopher eating
- P1–P5: chopsticks free

And the transitions:

- T1–T5: end of meditation and start of waiting for the chopsticks
- T6–T10: seizing of the left and right chopsticks and start of meal
- T11–T15: end of meal and start of meditation

Figure 5.10 shows one of the five submodels that make up the complete model. In this example, the meditation time follows a uniform distribution U(1.10) minutes and the meal time also follows a uniform distribution U(1.10). The total simulation time is 10,000 min.

The following objects are employed in each submodel:

- The Source class object *InicializacionFilosofo* creates just 1 temporary entity in the initial instant that philosopher f1 represents.
- The Server class object M1 that codes the meditation time (Fig. 5.11).
- The C1 object of the Server class that codes the wait to seize chopsticks P1 and P5, the seizing of the chopsticks, the meal time and the release of the chopsticks (Fig. 5.12). Simio forces the synchronization of the resources seize, in this case when both resources are requested from the same field in the "Secondary Resources".

The synchronous seize of both chopsticks guarantees that the process will not have blockages. Some results obtained from 10 replications are:

- Utilization of the chopsticks: 0.91
- Average time of waiting to have the two chopsticks: 0.11 min

Another variant of the same problem consists of the philosopher trying to reach first the chopstick on his left, and then, the chopstick on his right. He releases both chopsticks simultaneously once he has finished eating and begins to meditate. In this approximation, the blockage is possible when the five philosophers have a chopstick in their left hand. This can happen when all the philosophers want to start

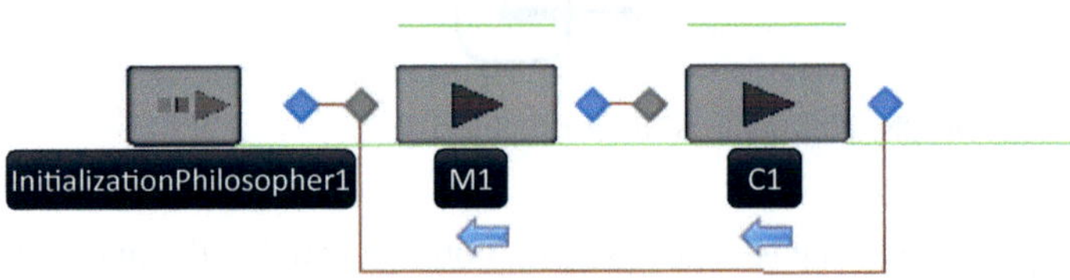

Fig. 5.10 Submodel in simio for the philosophers' dinner

Properties: M1 (Server)

Process Logic	
Capacity Type	Fixed
Initial Capacity	1
Ranking Rule	First In First Out
Dynamic Selectio...	None
TransferIn Time	0.0
Processing Time	**Random.Uniform(1,10)**

Fig. 5.11 Parameterization of the M1 object of the Server class

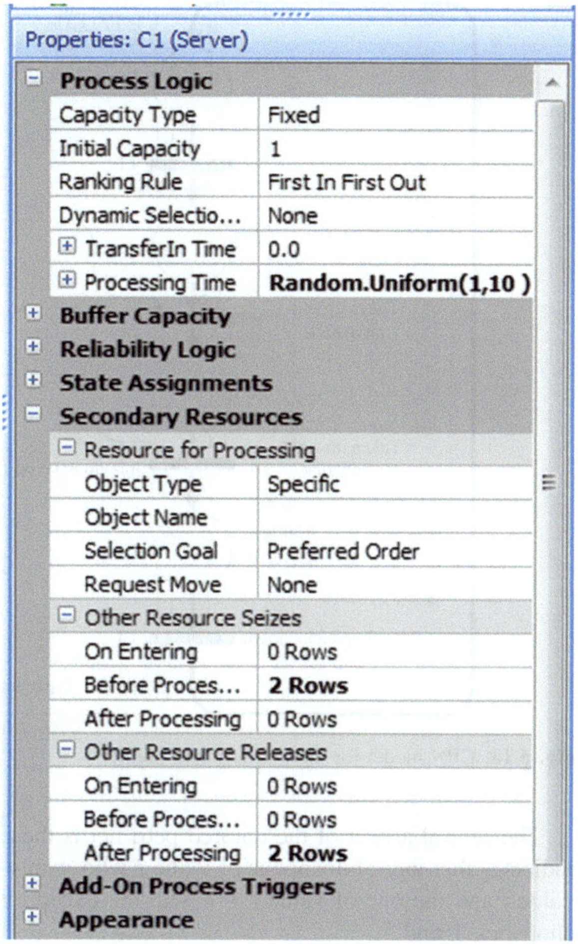

Fig. 5.12 Parameterization of the C1 object of the Server class

eating at the same time, take their left chopstick and then want to pick up the right one.

Unlike the previous Petri net, we can employ a much more compact representation by working with colored petri nets (RPC), as in Fig. 5.13. The statements for the colored petri net are:

- Color P = integer with 1…5
- Color F = integer with 1…5
- Attribute pi of color P
- Attribute pd of color P
- Attribute f of color F
- Previous function (f:F): pd = if f > 1 then f-1 else 5

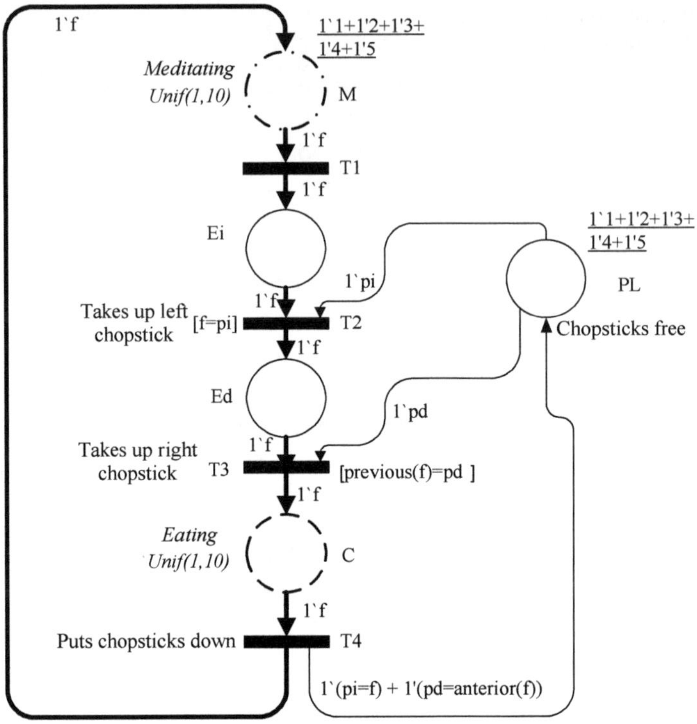

Fig. 5.13 CPN model for the philosophers' dinner

The central aspect of the colored petri net is the expression of guard T3, which indicates that any philosopher of value *f* who wants to eat, needs the chopstick of value *f* and the one of value $f + 1$ with the exception of philosopher 1 who needs chopstick 1 and 5.

Below is the formal specification of the above CPN:

$$\Sigma = \{P, F\}$$

$$P = \{M, Ei, Ed, PL\}$$

$$T = \{T1, T2, T3, T4\}$$

$$A = \{MaT1, T1aEi, EiaT2, PLaT2, T2AEd, EdaT3, PLaT3, T3aC, CaT4, T4aM, T4aPL\}$$

$$N(a) = (ORIGIN, DESTINATION)\ if\ a\ has\ an\ ORIGIN\ to\ DESTINATION\ format$$

$$C(p) = \begin{cases} P & \textit{if } p = PL \\ F & \textit{if } p \in \{M, Ei, Ed\} \end{cases}$$

$$G(t) = \begin{cases} f = pi & \textit{if } t = T2 \\ previous(f) = pd & \textit{if } t = T3 \\ true & \textit{if } t \in \{T1, T4\} \end{cases}$$

$$E(a) = \begin{cases} 1'f & \textit{if } a \in \{MaT1, T1aEi, EiaT2, T2aEd, EdaT3, T3aC, CaT4, T4aM\} \\ 1'pi & \textit{if } a \in \{PLaT2\} \\ 1'pd & \textit{if } a \in \{PLaT3\} \\ 1'(pi = f) + 1'(pd = previous(f)) & \textit{if } a \in \{T4aPL\} \end{cases}$$

Figure 5.14 shows the complete code at "Facility" level. The *In InitializationPhilosophers* object of the Source class creates five philosophers, and for each one of them, it initializes the value of the attribute for the philosopher's identifier (from 1 to 5). M1 is an object of the Server class with infinite capacity and a process time that follows the uniform distribution (1.10) min. Lastly, the object Ei_Ed_C of the Server class with infinite capacity codes the dinner time (uniform (1.10) min).

In the parameterization of the object Ei_Ed_C of the Server class (Fig. 5.15) we can see that we have the "Add-On Process Trigger" Ei_Ed_C_Processing that is executed before initiating the process and the Ei_Ed_C_Processed that is executed when the process is completed.

For each philosopher, the "Add-On Process Trigger" Ei_Ed_C_Processing calls up two consecutive Seize steps. The first being to seize the chopstick on the left and the second to seize the chopstick on the right. Note that the compacting of the Simio model in the "Facility" layer has been achieved by incrementing the complexity in the "Processes" layer. Also note, although it has no effect in this trigger, that the seize of multiple resources in a single Seize Step is not synchronic, Fig. 5.16.

Ei_Ed_C_Processed simultaneously frees the right and left chopsticks when the meal ends (Fig. 5.17).

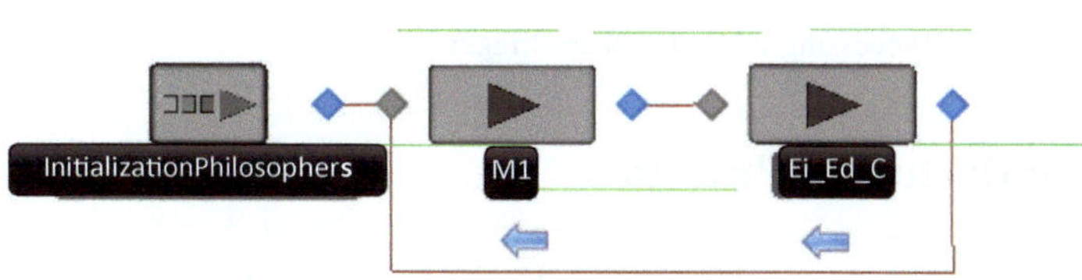

Fig. 5.14 Simio model for the philosophers' dinner

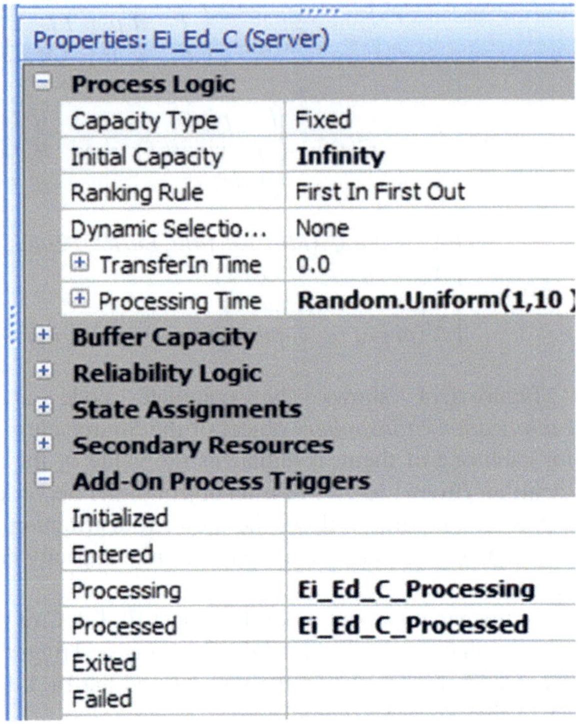

Fig. 5.15 Parameterization of the object Ei_Ed_C of the server class

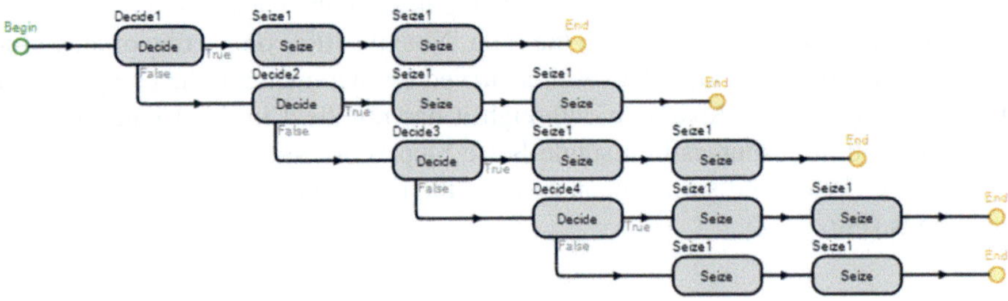

Fig. 5.16 Ei_Ed_C_Processing add-on process trigger

5.5 Manufacturing Process

This case is an adaptation of an example proposed by (Law 2000). Figure 5.18 shows an approach for the process involved. It consists of 5 work stations, one station for the entry/exit of pieces and one or more forklifts for transporting pieces between stations. Each work station has several identical machines and a queue without any size limit. The end goal of the study is to determine how many

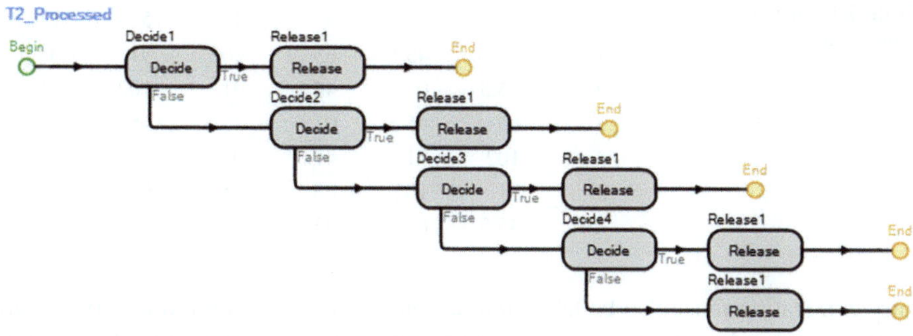

Fig. 5.17 Ei_Ed_C_Processed add-on process trigger

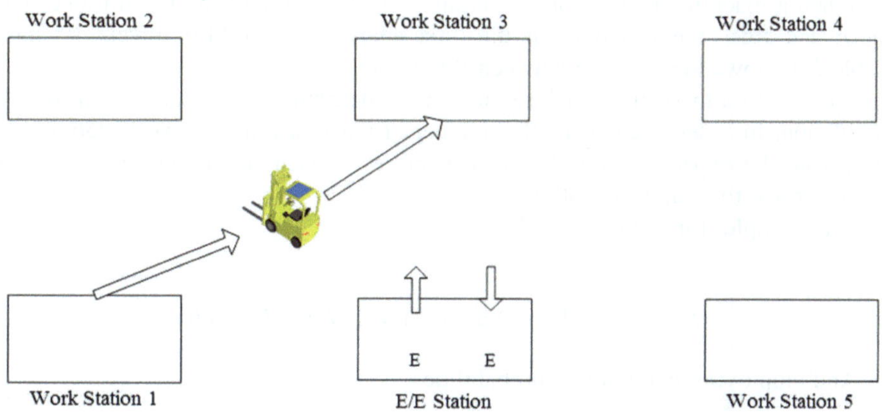

Fig. 5.18 Manufacturing process

Table 5.3 Circuits and mean service times

Type of piece	Circuit	Mean service time (hours)
1	3, 1, 2, 5	0.25, 0.15, 0.10, 0.30
2	4, 1, 3	0.15, 0.20, 0.30
3	2, 5, 1, 4, 3	0.15, 0.10, 0.35, 0.20, 0.20

machines are required in each station and how fast the forklift should be in order to appropriately meet the production specifications.

It is assumed that the pieces to be processed arrive at the entry/exit station depending on a mean exponential distribution of 1/15 h/piece. There are three types of pieces. The probability that a piece is type 1 is 0.3; type 2, 0.5 and type 3, 0.2. Each type of piece has a different circuit, exactly as represented in Table 5.3.

Given that the pieces enter and exit through the E/E station, the trips from the entry/exit station to the first work station and from the last work station to the

Table 5.4 Distance between stations, in meters

	1	2	3	4	5	E/E
1	0	45.5	65	102	91	45.5
2	45.5	0	45.5	91	102	65
3	65	45.5	0	45.5	65	45.5
4	102	91	45.5	0	45.5	65
5	91	102	65	45.5	0	45.5
E/E	45.5	65	45.5	65	45.5	0

entry/exit station must also be taken into account. The time required to process any piece in any machine follows a gamma distribution with a shape parameter of value 2 ($\alpha = 2$), whose mean depends on the type of piece and on the station being worked on. The above table also shows the mean work time per piece and station.

When a machine finalizes an operation, it is blocked until the forklift takes the piece and frees the machine. In this first analysis the forklift travels 1.5 m/s. Table 5.4 shows the distance between the different stations.

We can do a theoretical analysis in order to determine the initial design for the simulation. In order to calculate the number of machines in each work station, we start from the expected time of arrival of pieces at each station and the expected mean service time in each station.

For example, for station 2 we have

$$E(A) = (1/15)/(0.3 + 0.2) = 1/7.5 = 0.133 \text{ h}$$

And employing conditioned probabilities,

$$E(S) = \frac{0.3 \cdot 0.1 + 0.2 \cdot 0.15}{0.3 + 0.2} = 0.12 \text{ h}$$

Given that the utilization has to be under or equal to 1 we have

$$\rho = \frac{\lambda}{w \cdot s} = \frac{E(S)}{E(A) \cdot s} = \frac{0.12}{0.133 \cdot s} \leq 1$$

and

$$s \geq \frac{E(S)}{E(A)} = \frac{0.12}{0.133} = 0.9 \Rightarrow 1$$

The number of theoretical machines per station is summarized in Table 5.5.

A similar calculation can be done to evaluate if the forklift has sufficient capacity. Given that the pieces are type one with a probability of 0.3, the expected time between arrivals $E(A)$ = (1/15)/0.3 = 0.222 h. The service time of the forklift is the travel time associated with the *E/E* circuit, 3, 1, 2, 5 and *E/E*

Table 5.5 Number of machines required in each station

Work station	$E(A)$ (hours/work)	$E(S)$ (hours/work) machine	Number of machines
1	0.066	0.215	3.25 ⇒ 4
2	0.133	0.120	0.9 ⇒ 1
3	0.067	0.265	3.95 ⇒ 4
4	0.095	0.165	1.73 ⇒ 2
5	0.133	0.220	1.65 ⇒ 2

Table 5.6 Number of forklifts needed

Type of piece	$E(A)$ hours/work	$E(S)$ (hours/work)	Number of forklifts
1	0.222	0.056	0.25
2	0.133	0.051	0.38
3	0.333	0.084	0.25
All the pieces			**0.88** ⇒ 1

$$E(S) = (45.5 + 65 + 45.5 + 102 + 45.5)\ \mathrm{m}/1.5\ \mathrm{m/s}\ *\ 1/3600\ \mathrm{h/s} = 0.056\ \mathrm{h}$$

and the number of forklifts needed for the type 1 pieces is

$$s = \frac{E(S)}{E(A)} = \frac{0.056}{0.222} = 0.25$$

The number of forklifts needed is summarized in Table 5.6.

These theoretical calculations suffer from two major defects:

1. They do not take into account the proportion of time when the machines are blocked.
2. They do not take into account the movements of the forklifts without a load.

Thus, it is advisable to do the simulation study in order to improve the theoretical design. Figure 5.19 shows the partial Petri net of the system. We can see that the machine ends the process and is blocked while waiting for the forklift. Place P4 considers the time it takes the forklift to move from where it is at that instant to station 2, and the time it takes it to move from station 2 to 5.

The places are:

- P1: pieces waiting to be processed in station 2
- P2: piece being processed in station 2
- P3: piece waiting for the forklift. The machine is blocked
- P4: movement of the forklift from where it is to station 2
- P5: transport to station 5
- P6: pieces waiting to be processed in station 5

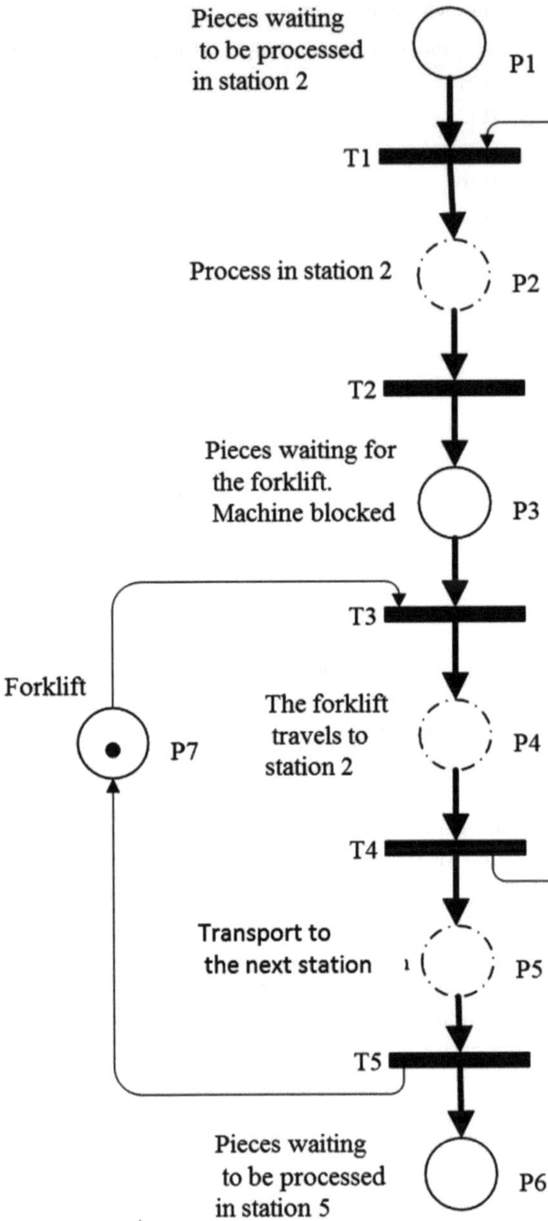

Fig. 5.19 Partial petri net for the manufacturing system

And the transitions,

- T1: seize of the machine of station 2 and start of process
- T2: end of process and start of blocking

- T3: seize of forklift
- T4: arrival at station 2 and release of machine
- T4: end of transport and release of R1 robot

In this example we opted not to get the Petri net of the complete system. In general, we use these nets as a support for modeling and later coding. The decision about what parts are to be modeled in Petri nets and the level of detail of the net itself is a choice that has to be made.

The Simio model is given in Fig. 5.20. It consists of a set of objects of the Server class (*W*1, *W*2, *W*3, *W*4, *W*5), an object of the Source class (*arrival*) and an object of the Sink class (*exit*), connected to each other by a transport network formed in its central rectangular zone by objects of the Path class whereby the *Forklift* object of the Vehicle class is moved. The entry (or exit from each work station *Wi*) TranferNodes are connected by Connectors to the TranferNode of the transport network. For example, the TransferNode *TN1o* is connected to the TransferNode *Input@Wi* and the TranferNode *TN1i* is connected to the TranferNode *Output@W1*.

For every arrival of a piece, the Source1_CreatedEntity Add-On Process Trigger is executed as shown in Fig. 5.21. The Decide Steps bifurcate the token along three paths with probabilities 0.3, 0.5 and 0.2.

The Steps Set Table specifies the sequence that the pieces have to follow according to their type. For example, Fig. 5.22 has defined the sequence for the type 1 piece and is formed by the set of TranferNodes to be visited. The Steps

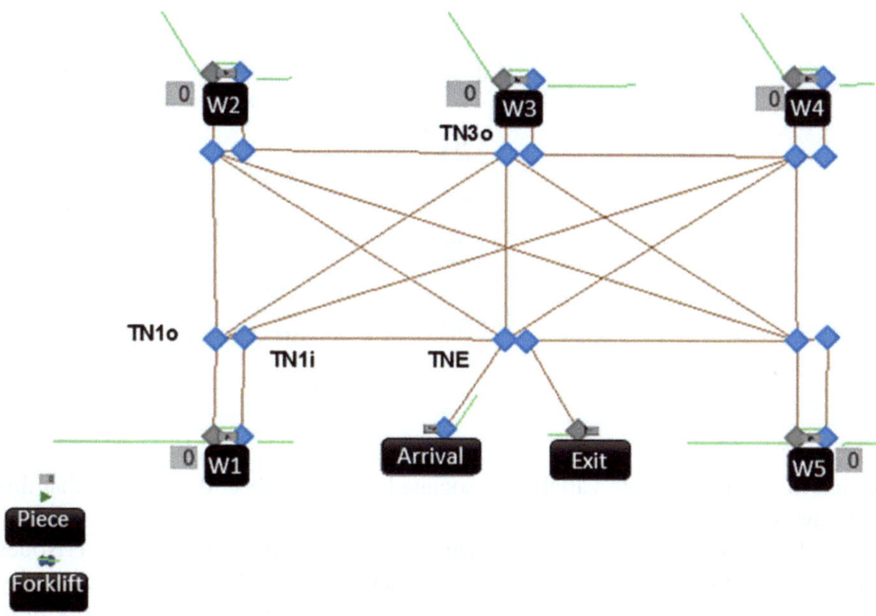

Fig. 5.20 Simio model for the philosophers' dinner

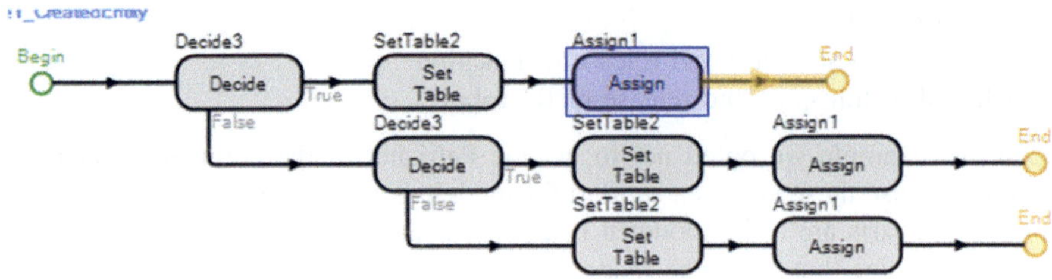

Fig. 5.21 Source1_CreatedEntity add-on process trigger

Fig. 5.22 Sequence of type 1 pieces

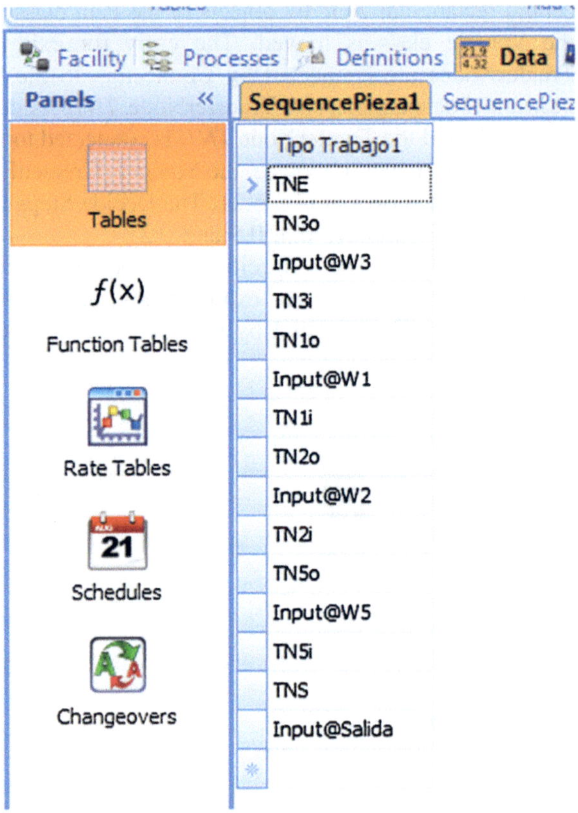

Assign memorize, in state variables associated with the piece, the runtime that this will have in each work station.

The last aspect to be highlighted is the parameterization of the TransferNodes (Fig. 5.23). In the OutputLlegada node we specify that the Destination of the entity (piece) is chosen depending on its sequence. In the TNE node we ask the forklift to go to the next Destination. In the TN3o node, the entity frees the forklift and

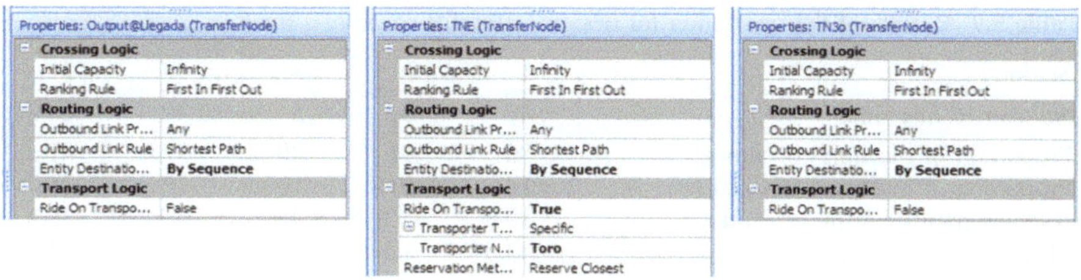

Fig. 5.23 Parameterization of the "transfernode"s Ouput@Llegada, TNE, TN3o

Table 5.7 Results of the simulation for design 1

Station	1	2	3	4	5
Number of machines	4	1	4	2	2
Proportion of machines occupied	70.36	67.29	83	72.61	60
Proportion of machines blocked	21.16	32.51	16.2	27	23
Average number of pieces in queue	2.79	295.2	199.8	179	1.29
Maximum number of pieces in queue	18	487	335.3	309.5	12
Daily average production	92.7				
Average time in the system	44.4				
Proportion of loaded forklift movement	75				
Proportion of unloaded forklift movement	24				

continues with the sequence that takes it to the Input@W3 node, for entry to the work station 3.

Table 5.7 shows the results of the simulation for the theoretical design. The study was done with 10 replications, 320 h of simulation and a warm-up time of 64 h. The daily production achieved with a value 92.7 pieces is far from the target production of 120. We can also see the high number of average and maximum pieces in the queue at stations 2, 3 and 4.

In design 2 a machine is added to stations 2, 3 and 4. The results (Table 5.8) are better, but the daily target production is not yet reached. One aspect that seems important is the high proportion of blocked machines. This is due to the time the

Table 5.8 Results of the simulation for design 2

Station	1	2	3	4	5
Number of machines	4	2	5	3	2
Proportion of machines occupied	74	43	75	54	69
Proportion of machines blocked	25	37	22	32	30
Average number of pieces in queue	138.3	1	21	2.2	163.3
Maximum number of pieces in queue	244.7	9.8	52	16.2	266.4
Daily average production	107.4				
Average time in the system	21.8				
Proportion of loaded forklift movement	83				
Proportion of unloaded forklift movement	16				

Table 5.9 Results of the simulation for design 3 (forklift speed at 2 m/s)

Station	1	2	3	4	5
Number of machines	4	2	5	3	2
Proportion of machines occupied	80	44	80	57	81
Proportion of machines blocked	14	10.8	10	15	16
Average number of pieces in queue	10.8	0.3	4.4	0.7	28.2
Maximum number of pieces in queue	43.5	7.3	27.1	10.8	62.7
Daily average production	120				
Average time in the system	3.9				
Proportion of loaded forklift movement	66				
Proportion of unloaded forklift movement	27				

forklift takes from when the request is made to extract the manufactured piece from the machine until the forklift loads it.

One way of making an improvement may be to increase the speed of the forklift from 1.5 to 2 m/s. The results of Table 5.9 show that the improvement is notable, with the daily target production being achieved. In the event that the maximum number of pieces in stations 1 and 5 is excessive, the possibility of incorporating another machine into stations 1 and 5 could be raised.

5.6 Automated Warehouse

This example is a simplification of the problem proposed in (Guasch et al. 2011). The system to be studied is an intermediate warehouse for paper spools managed by two pallet elevators that automatically feed several rotating pieces for printing newspapers. Figure 5.24 shows a diagram of the warehouse made up by a central rail along which the two pallet elevators move; 88 warehousing positions in the horizontal direction of the rail.

For each horizontal position we have, in general, 6 warehousing positions; 3 on both sides of the pallet elevator occupying three different positions in the vertical direction, see Fig. 5.25.

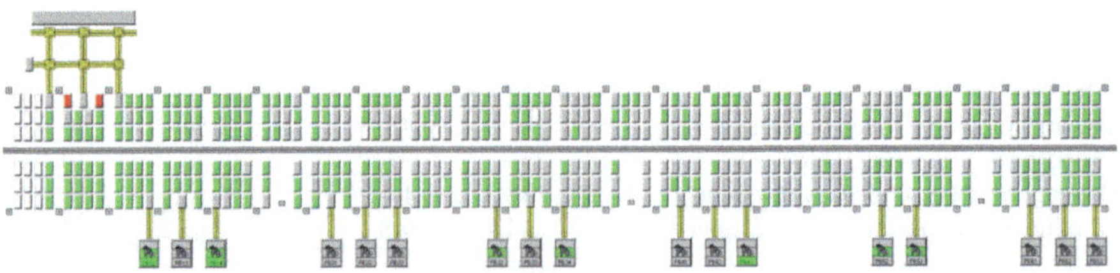

Fig. 5.24 Diagram of automated warehouse

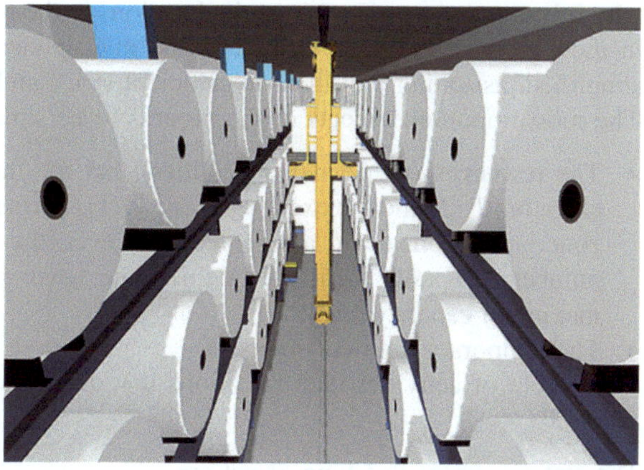

Fig. 5.25 View of the warehouse from the central rail

Given that the two pallet elevators circulate along the same central rail, the objective of the study is to determine the work zone assigned to each pallet elevator in order to distribute the work load equitably between the two. The working configuration using two pallet elevators is necessary given that just one is not enough to cover the work demand during the peak hours from 0 to 2 a.m.

Briefly, pallet elevator 1 works in the left zone; 2 in the right zone, and a block of 2 horizontal positions (12 warehousing positions) is reserved as a zone exclusively for the exchange of spools between the pallet elevators. To avoid the two pallet elevators hitting each other, the control system only allows the entry of one pallet elevator at a time into the exchange zone.

The travel time between two horizontal positions depends on the distance that is run in horizontal positions (*dh*)

$$Time = (dh - 5) * 0.97 + 15.7\ (\mathrm{s})$$

If $dh = 0$, the time is also 0. The travel time between two vertical positions depends on the distance that is run in vertical positions (*dv*) and whether it is up or down (Table 5.10). Since horizontal and vertical movements are performed in parallel, the movement time between two different positions is the maximum of the vertical and horizontal times.

The time taken to transfer a spool from the warehouse to the pallet elevator or from the pallet elevator to the warehouse is 21 s.

To do the study, we start from a file of real orders in the period from 0 to 2 a.m. during 7 consecutive days. These orders, although given in the working configuration with a single pallet elevator, are independent of the way of working. This is

Table 5.10 Vertical travel times

Dv	Direction	Time (s)
1	Up	10.2
2	Up	15
1	Down	12.2
2	Down	23

why it is unnecessary to manage the locations in the model, except for the locations in the exchange zone. Nevertheless, the management of said locations has been simplified, assuming that there is sufficient space and mean access times.
The most important aspects of the process to be simulated are:

- The resources (or permanent entities) that are involved in the process: pallet elevators T (2) and the exchange zone E (1). The set of locations in the exchange zone is not considered a resource. However, the simulation shows us that the number of spools there are in this zone never at the same time exceeds its maximum capacity.
- The temporary entities: orders O.
- The Set of activities and transitions that are made explicit in the colored petri net.

The statements for this type of net are:

- Color T: integer with 1…2
- Color H: integer with 1…88
- Color E: exchange zone
- Color O = HxH: order with a position of origin and a position of destination
- Attribute t of color T
- Attribute ha, ho, hd of color H, where ha is the current horizontal position of the pallet elevator, ho the position of origin where the spool to be picked up is to be found and hd the destination position of the spool to be placed.
- Attribute e of color E
- Function z(h:H): if h <=hi then 1 else 2

Figure 5.26 shows the Petri net of the system. The places are:

- P1: transport order with position of origin ho and destination hd
- P2: transport order awaiting pallet elevator 1 that is in the horizontal position ha
- P3: pallet elevator 1 moves from ha to ho in order to load the spool and from ho to hd in order to unload it
- P4 and P5: the same as P2 and P3, but for orders that only specify pallet elevator 2
- P6: the two free pallet elevators
- P7: the free exchange zone
- P8: order waiting for pallet elevator 1
- P9: pallet elevator 1 moves from ha to ho in order to load the spool and from ho to position hi-1 waiting to be able to enter the exchange zone
- P10: pallet elevator 1 waits for the exchange zone to be free
- P11: movement of pallet elevator 1 to the exchange zone in order to leave the spool and withdraw to position hi-1
- P12: waiting for pallet elevator 2 to be free
- P13: movement of pallet elevator 2 to position hi + 2, bordering on the exchange zone
- P14: pallet elevator 2 waits for the exchange zone to be free

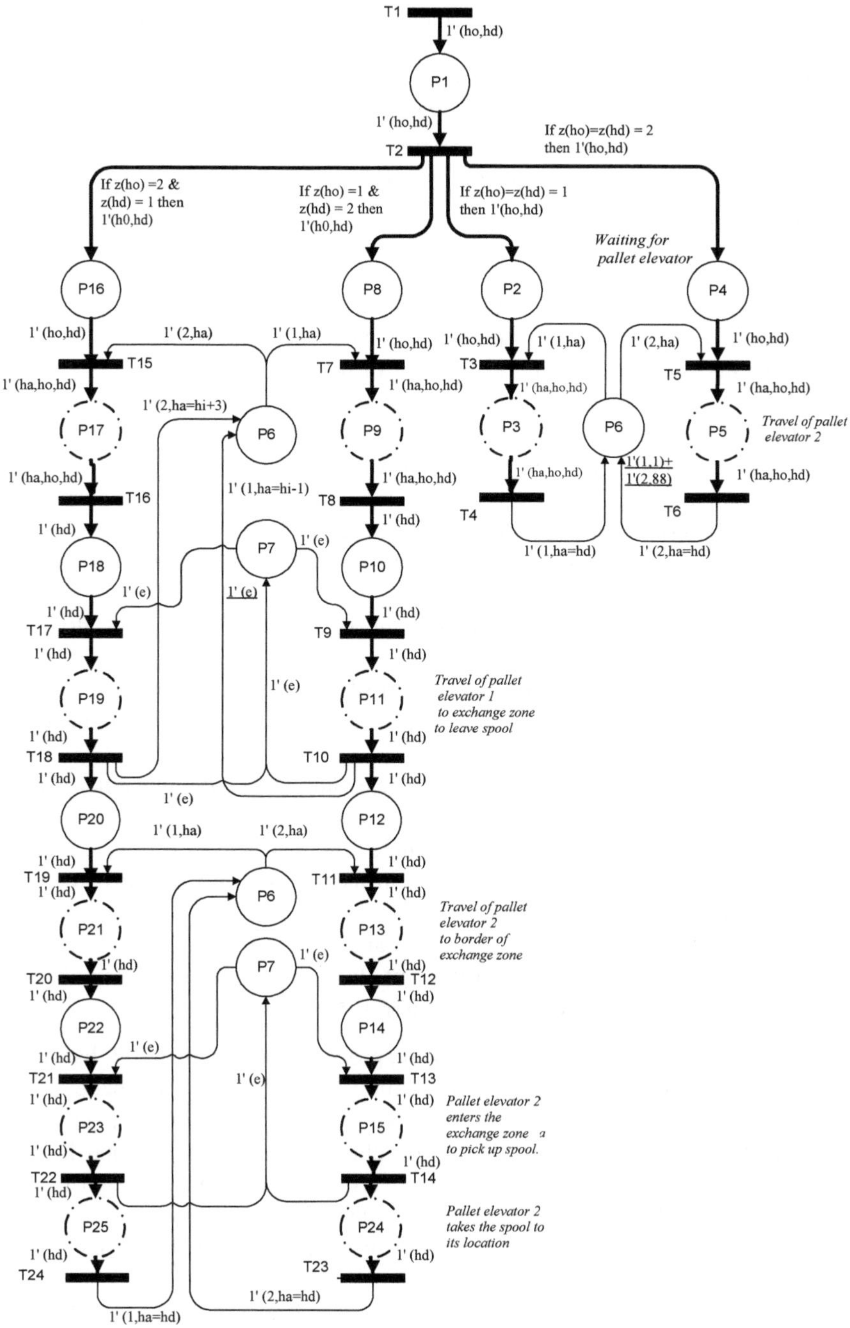

Fig. 5.26 Colored petri net of the model

- P15: pallet elevator 2 enters the exchange zone in order to pick up the spool and leave this zone
- P23: pallet elevator 2 takes the spool to its final location
- P16, P17, P18, P19, P20, P21, P22, P23 and P25: the same as P8, P9, P10, P11, P12, P13, P14, P15 and P24 but the first stage of the process is done by pallet elevator 2 and the second, pallet elevator 1.

And the transitions:

- T1: arrival of orders.
- T2: distribution of orders as a function of whether we only need pallet elevator 1 or 2, or a combination of both when the exchange zone is used.
- T3: seizing of pallet elevator 1 and start of movement.
- T4: end of operation and release of pallet elevator 1.
- T5 and T6: the same as T3 and T4, but for pallet elevator 2.
- T7: seizing of pallet elevator 1 and start of movement.
- T8: the pallet elevator has arrived at the position hi-1 before the exchange zone and stops, waiting for this zone to become free.
- T9: the exchange zone is seized and the movement starts in order to leave the spool in the exchange zone.
- T10: pallet elevator 1 has completed the operation and remains stopped in position hi-1.
- T11: seizing of pallet elevator 2 and start of movement.
- T12: pallet elevator 2 has reached the position hi + 2 waiting for the exchange zone to become free.
- T13: seizing of the exchange zone. Pallet elevator 2 starts the movement of entry to this zone.
- T14: pallet elevator 2 leaves the exchange zone and leaves it free.
- T23: pallet elevator 2 leaves the spool in its location and remains free in this position waiting for new orders.
- T15, T16, T17, T18, T19, T20, T21, T22 and T24: the same as T7, T8, T9, T10, T11, T12, T13, T14 and T24 but the first stage of the process is done by pallet elevator 2 and the second, pallet elevator 1.

Figure 5.27 shows the Simio code for the process to be simulated. It has a direct parallelism with the colored petri net. The initial stage consists of the pallet elevator going in empty to the position where the spool is and later moving on loaded. It has been coded with two objects of the Server class, for example P5a and P5b. The coded model includes both horizontal and vertical movement, although the CPN does not explain the attributes associated with the vertical positions. The times associated with each one of the different movements is calculated in the Add-On Process Triggers for the different objects of the Server class.

Table 5.11 shows the results for each one of the simulations performed. We can see that the exchange position 45 (and 46) is the one that has associated a shorter mean time in the system. This is the time that elapsed between order being received to transfer a spool and the order being executed. For this specific case, the

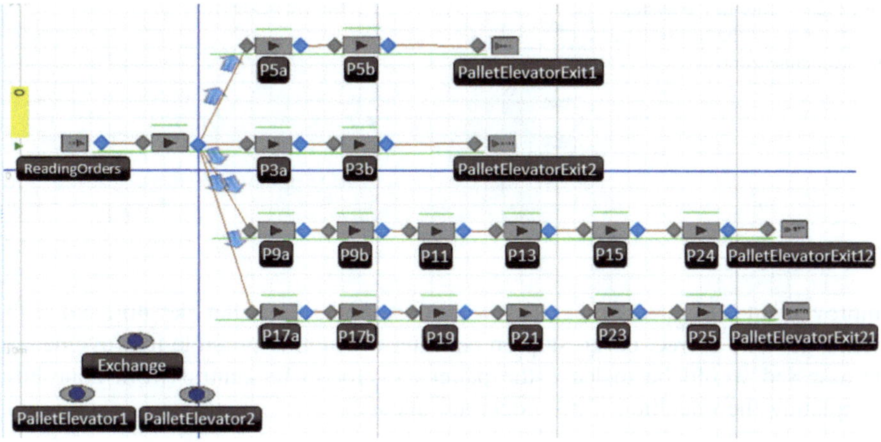

Fig. 5.27 Simio code for the warehouse model

Table 5.11 Results of the simulation study

Exchange position	Utilization tr1	Utilization tr2	Waiting time	Maximum wait	Time in system	Exchange factor
0	0	0.76	247	918	351	0
10	0.18	0.79	294	1373	472	0.21
20	0.25	0.73	205	1268	374	0.25
30	0.4	0.63	82	679	249	0.3
35	0.49	0.51	64	541	214	0.29
40	0.54	0.43	57	491	192	0.24
45	0.55	0.37	58	498	188	0.21
50	0.59	0.35	85	913	216	0.23
60	0.66	0.27	108	763	237	0.2
70	0.78	0.14	263	1107	406	0.17
80	0.78	0.11	289	1253	422	0.12
89	0.76	0	247	918	351	0

utilization of the two pallet elevators is 0.55 and 0.37 respectively; the mean queuing time of the orders is 58 s; the order that waited longest has taken 498 s to be served and 21% of the operations have employed the exchange zone.

Figure 5.28 shows the mean time of the orders in the system as a function of the position of the exchange zone.

The model coded in Simio is a simplification of the reality and it is possible to get better results with other management policies. In this model it is assumed that before entering the exchange zone we stop the pallet elevator in order to request permission to enter and on exiting we stop the pallet elevator in order to free the exchange zone. These stops have a significant associated loss of time. An

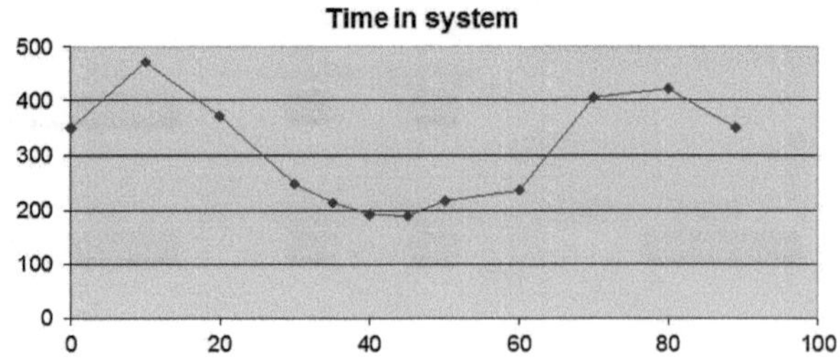

Fig. 5.28 Time in the system as a function of the position of the exchange zone

improvement is to release on exiting without stopping the pallet elevator, and, if the exchange zone is free, not to stop on entering it. Another possible improvement to be assessed would be to leave the pallet elevator in its central work zone after completing the operation if no orders are queued.

References

Dijkstra, E. W. (1971). Hierarchical ordering of sequential processes. *Acta Informatica, 1,* 115–138.

Guasch, A., Piera, M. A., & Figueras, J. (2011). Automatic warehouse modeling and simulation. *International Journal of Simulation and Process Modeling*, *6*, 288–296. ISSN: 1740-2123.

Law, A. M., & Kelton, W. D. (2000). *Simulation modeling and analysis*. New York: McGraw-Hill.

MIX
Papier aus verantwortungsvollen Quellen
Paper from responsible sources
FSC® C105338

If you have any concerns about our products,
you can contact us on
ProductSafety@springernature.com

In case Publisher is established outside the EU,
the EU authorized representative is:
Springer Nature Customer Service Center GmbH
Europaplatz 3, 69115 Heidelberg, Germany

Printed by Libri Plureos GmbH
in Hamburg, Germany